E-MRS 2011 Fall Symposium K: Solution-Derived Electronic-Oxide Films, Nanostructures and Patterning, from Materials to Devices

IOP Conference Series: Materials Science and Engineering
Volume 30

Warsaw, Poland
19 - 23 September 2011

ISBN: 978-1-63439-841-1
ISSN: 1757-8981

Printed from e-media with permission by:

Curran Associates, Inc.
57 Morehouse Lane
Red Hook, NY 12571

proceedings.com

Some format issues inherent in the e-media version may also appear in this print version.

Printed by Curran Associates, Inc. (2015)

Institute of Physics
Dirac House, Temple Back
Bristol BS1 6BE UK

Phone: 44 1 17 929 7481
Fax: 44 1 17 920 0979

techtracking@iop.org

Additional copies of this publication are available from:

Curran Associates, Inc.
57 Morehouse Lane
Red Hook, NY 12571 USA
Phone: 845-758-0400
Fax: 845-758-2634
Email: curran@proceedings.com
Web: www.proceedings.com

Table of contents

Volume 30

E-MRS 2011 Fall Symposium K: Solution-Derived Electronic-Oxide Films, Nanostructures and Patterning, from Materials to Devices
19–23 September 2011, Warsaw, Poland

Accepted papers received: 16 January 2012
Published online: 15 February 2012

Papers

012001

Solution-derived photocatalytic films for environmental cleaning applications U L Štangar, M Kete, A Šuligoj and M Tasbihi pg. 1

012002

Low temperature processing of alkoxide-derived PMN thin films T Ohno, Y Gotoh, N Sakamoto, N Wakiya, T Kiguchi, T Matsuda and H Suzuki pg. 8

012003

Mesoporous sol-gel deposited SiO_2-SnO_2 nanocomposite thin films A A Ponomareva, V A Moshnikov and G Suchaneck pg. 14

012004

Bi-Sr-Co-O thermoelectrics prepared by sol-gel methods with modified gel decomposition K Rubešová, T Hlásek, V Jakeš, D Sedmidubský and J Hejtmánek pg. 19

012005

Influence of substrate bias voltage on the physical, electrical and dielectric properties of RF magnetron sputtered TiO_2 films P Kondaiah, M C Sekhar, S V Jagadeesh Chandra, R Martins, S Uthanna and E Elangovan pg. 24

PREFACE

The European Materials Research Society (EMRS) Fall Meeting 2011 Symposium K: Solution-Derived Electronic-Oxide Films, Nanostructures and Patterning, from Materials to Devices brought together the scientific community working in the field of solution-derived electronic-oxide thin films, patterned structures and related devices. The meeting took place on 19–23 September 2011 in Warsaw, Poland. The participants shared their results in 12 sessions and 2 poster sessions. There were 15 invited talks, 31 oral presentations and 28 poster contributions, 74 in total.

The meeting provided an excellent opportunity for an overview and a forum for the interchange of information and expertise on the cutting-edge research, advanced technologies and applications in the field.

The symposium addressed solution processing of different materials with specific functionalities, for example transparent conductive oxides which can be used in amorphous phase, and can therefore be processed at temperatures as low as 200°C, or thin film materials for solid oxide fuel cells, which require high processing temperatures to obtain the required properties. In addition to already established thin film processing, novel direct patterning methods were presented. The influence of processing on materials and functional properties of thin films and structures and integration issues were addressed.

We would like to thank all the oral and poster contributors, and the session chair-persons, who devoted their time and energy to the success of this symposium. We are indebted to the Advisory Board for their advice. We would like to thank the European Materials Research Society for the support and organisation of the meeting.

Barbara Malic
Viorica Musat
Marlies Van Bael
Theodor Schneller
January 2012

IOP Conf. Series: Materials Science and Engineering **30** (2012) 012001 doi:10.1088/1757-899X/30/1/012001

Solution-derived photocatalytic films for environmental cleaning applications

U Lavrenčič Štangar[1], **M Kete, A Šuligoj and M Tasbihi**
Laboratory for Environmental Research, University of Nova Gorica, Vipavska 13, SI-5001 Nova Gorica, Slovenia

E-mail: urska.lavrencic@ung.si

Abstract. When photocatalytic water treatment is concerned, suspended catalyst in the aqueous phase is usually more efficient than immobilized on an inert support, but in the former case an undesirable separation/recycling step is needed. We have therefore concentrated on the preparation of immobilized catalysts in the form of films on glass and aluminium supports. The low-temperature sol-gel processing route to obtain transparent thin TiO_2/SiO_2 films for self-cleaning purposes and thicker TiO_2/SiO_2 coatings for efficient removal of pollutants in water and air are presented. The synthesis is based on a production of a nanocrystalline titania sol with a silica binder that after deposition does not require thermal treatment at high temperatures. Depending on the target application, some specific synthesis parameters and photocatalytic activity testing conditions are illustrated. For water-cleaning coatings fast kinetics is required, which was achieved by addition of a highly active titania powder into the sol. The same preparation procedure was used to prepare efficient air-cleaning coatings. On the other hand, self-cleaning films were thinner and transparent to keep the original appearance of the substrate and they solidified at ambient conditions. Advanced methodologies to evaluate photocatalytic activity of the films were applied.

1. Introduction

Heterogeneous photocatalysis, in which different semiconductor oxides are used, is a fast growing and perspective area of research with a variety of applications, including water cleaning, air remediation and self-cleaning surfaces. Titanium dioxide in the form of powder dispersions and thin nanostructured films continues to be the most popular photocatalyst for the purification of air and water. When exposed to UVA light it has a high oxidizing power that can convert harmful organic contaminants into carbon dioxide, water and mineral acids via the complete mineralization process (reviews [1-4]). If TiO_2 is immobilized on a rigid support as a thin (nano- or micro-scale) film, a problem of final particle-fluid separation for the catalyst recycling in slurry-type photoreactors is avoided. A large number of studies have been made to immobilize the catalyst on a rigid support via different deposition techniques, such as sputtering, chemical vapour deposition, thermal oxidation, sol-gel technique etc. [1]. Sol-gel techniques offer some advantages compared to other solution methods or gas phase deposition techniques, such as ease and versatility of processing, homogeneity at a molecular level, as well as mild and ambient atmosphere reaction conditions. However, there are certain problems regarding the photocatalytic activity of the immobilised TiO_2, *i.e.* (i) a lower rate of electron and hole transfer across the semiconductor-solution interface because of the decreased

[1] To whom any correspondence should be addressed.

catalyst surface in immobilized TiO_2; (ii) light absorption characteristics because of the limited amount of TiO_2 particles capable of absorbing the light; (iii) a substrate since it may have a negative influence on the photocatalytic activities of the titania films, *e.g.* sodium diffusion from glass into the interior of the films during calcination. The latter is avoided by using the low-temperature processing route [5], where crystallization is promoted already in solution and has therefore a higher application value.

Although TiO_2 photocatalysis was originally more oriented toward its use in water treatment processes, it turned out that self-cleaning surfaces are more successful from application point of view than water or air remediation because of the two main reasons [3]. First, the fast kinetics of photocatalytic degradation is not absolutely required in this case and second, high quantum yields of photoinduced superhydrophilicity, which is the main contributor to the self-cleaning effect, are achieved even at low illumination levels. In spite of emerging number of self-cleaning photocatalytic products in the market, there is still a lack of ideal standard method for determining self-cleaning efficiency. Some new methodologies for evaluating self-cleaning and photocatalytic activities [6, 7] are applied herein for the two types of coatings that were prepared by the low-temperature sol-gel processing route: (i) TiO_2/SiO_2 thin transparent films for self-cleaning purposes, and (ii) TiO_2/SiO_2 thicker opaque films for organics removal from water and air.

2. Experimental

2.1. Preparation of self-cleaning TiO_2/SiO_2 thin films

The preparation process relies on the patented procedure [8], which comprises the preparation of an acidic aqueous sol containing photochemically active TiO_2 nanoparticles (in this work we used an acidic aqueous suspension with TiO_2 concentration 100 g/L provided by Cinkarna Celje [9]), the addition of a silica binder solution and the dilution of the prepared sol with water and organic solvents (1-propanol, 2-propoxyethanol). The silica binder solution was prepared from tetraethyl orthosilicate (TEOS), colloidal SiO_2 Levasil 200/30% aqueous solution, HCl to catalyse TEOS hydrolysis and 1-propanol. A final coating solution contained 2.1 wt.% of TiO_2 with the molar ratio n(Ti) : n(Si) = 1 : 1. It was deposited on clean glass surfaces either by dip-coating (10 cm/min, microscope objective glass slides) or by spraying on bigger glass surfaces (25 cm × 25 cm). The resulting coatings (samples UNG10-7Cdip and UNG10-7Cspray, resp.) were of a good optical quality, completely transparent for visible light and they did not require any thermal treatment (they solidify under sun or over radiator for a day and retain their mechanical stability for years).

2.2. Preparation of TiO_2/SiO_2 films for organics removal from water and air

TiO_2/SiO_2 composites were synthesized and immobilized on Al-sheets using two commercial titania powders (Degussa P25 and Millennium PC500, mass ratio 1:1) with a silica/titania binder and without a binder, thoroughly described in [10]. Silica/titania binder was prepared from the following main constituents: refluxed titania aqueous sol derived from titanium isopropoxide precursor, tetraethyl orthosilicate and commercial colloidal silica Levasil 200/30%. The source of highly photocatalytically active TiO_2 was mixture of Degussa P25 and Millennium PC500 suspended in the binder solution and ultrasonically treated. Before applying this hybrid sol on the Al-sheets (24 cm × 1.2 cm for aqueous photocatalysis and 34 cm × 1 cm for gaseous photocatalysis), they were protected by the silica thin film derived from trimethoxy silane, tetraethyl orthosilicate and Levasil 200/30% (sample Al-p). After each application of the sol, the resulting TiO_2/SiO_2 coatings were heat-treated at 150 °C for 1 h. The procedure was repeated until the amount of catalyst reached 1 mg/cm^2 (sample titania/binder/Al-p). As a reference photocatalyst, suspension of the two commercial titania powders without a binder component was applied on the Al-sheets (sample titania/Al).

2.3. Instrumental

The X-ray powder diffraction (XRD) patterns were obtained on a PANalytical X-ray diffractometer using CuKα1 radiation (1.5406 Å). Morphology of samples was evaluated by scanning electron microscopy (SEM) on a Zeiss SupraTM 3VP microscope (self-cleaning films) at the National Institute of Chemistry, Ljubljana and on a FEI-Sirion microscope (water- and air-cleaning films) at the Indian Institute of Science, Bangalore.

2.4. Determination of self-cleaning efficiency

Self-cleaning efficiency was determined by: (i) a conventional method following photodegradation of a solid fatty deposit over catalyst layer, and (ii) a novel testing method following photooxidation of an entrapped sodium salt of terephthalic acid (TPA) in a solid polymer matrix deposited over catalyst layer.

(i) *Contact Angle (CA) method.* As a model fatty contaminant we used methyl stearate. The pure photocatalyst films were preirradiated in the photochamber (type of the lamps: CLEO 20 W, 438 mm x 26 mm, Philips; broad maximum at 355 nm, irradiation intensity 2.0-2.2 mW/cm^2) for 24 hours. After the pre-irradiation, the methyl stearate coatings were prepared from a solution of methyl stearate in n-hexane (0.2 M) by dip-coating with a pulling speed of 20 $cm \cdot min^{-1}$. Then the films were positioned back in the photochamber on places with the same UVA intensities and were irradiated for a certain time in air at 25 °C. The degradation of methyl stearate was studied by measuring the water contact angles (CA) at room temperature using a horizontal microscope with a protractor eyepiece (Contact Angle Meter (CAM-100), KSV Instrument, Ltd. Finland). Specimens were taken out of the illumination chamber for CA measurements after irradiation for different times. The data reported herein correspond to the stable value of the angles obtained by averaging three repeated measurements.

(ii) *Fluorescence method.* It is based on the homogeneous entrapment of a sodium salt of terephthalic acid (TPA) in the polymer host (hydroxyethyl-cellulose) to form a transparent thin solid layer over photocatalyst surface and on highly sensitive fluorescence detection of the one of TPA oxidation products (hydroxyterephthalic acid = HTPA) [6]. The thickness of the transparent TPA coating obtained is about 1 μm. The pure photocatalyst films were pre-irradiated in the photochamber described above before TPA coating was deposited. Then the samples were irradiated for a certain time. For analysis by HPLC (HP 1100 Series chromatograph, coupled with a FLD detector), the TPA layer was washed with a mixture of ethanol/water (1:1 volume ratio). A fitting of the experimental data obtained after different irradiation times was performed according to a simplified kinetic analysis for the initial degradation (till the plateau). The degradation of TPA followed a pseudo-zero order kinetics and consequently also the formation of HTPA followed a zero-order kinetics (formation rate constant = k_1). The seconds step is the degradation of HTPA to final inorganic products, which was no longer quantitatively followed.

2.5. Determination of photocatalytic activity in water

The photocatalytic activity of the as-prepared films in aqueous solution purged with oxygen was evaluated in a tailor-made chamber photoreactor using 6 low-pressure mercury fluorescent lamps as a UVA radiation source (CLEO 20 W, 438 mm × 26 mm, Philips; broad maximum at 355 nm). The photocatalytic cell is placed in the center of the reactor and it contains a teflon holder for up to 12 supported catalyst slides to be set around its axis [11]. 11 Al-slides with TiO_2/SiO_2 catalyst were used in this study. Aqueous solution (250 mL) of phenol (50 mg/L) acidified with perchloric acid to pH = 3.2 was used as a photodegradation medium. HPLC analyses in the dark phase of the experiment and after different irradiation times were performed on an HP 1100 Series chromatograph, while determination of total organic carbon (TOC) was done on Analytik Jena AG MULTI N/C 3100 apparatus.

2.6. Determination of photocatalytic activity in air

The photoreactor system for *in-situ* GC-MS photocatalytic experiments was recently constructed and optimized for powder samples [7]. The reactor chamber consists of 6 low-pressure mercury fluorescent lamps (Philips CLEO 15 W, broad maximum at 355 nm) and the reflective surface of polished aluminium placed behind the lamps. The reaction cell with immobilized catalyst on Al-sheets, fastened around the axis in a special teflon holder, was made from a Duran glass tube (42 mm inner diameter, 42 cm height) and positioned horizontally in the centre of the photoreactor [12]. Three Al-sheets with immobilized catalyst were used in the experiments. The regulated feed gas consisted of dry air, humidified air and toluene. The flow rates of both dry and humidified air were adjusted to 0.2 L/min to obtain the total air flow 0.4 L/min containing 45-50% humidity. Toluene was injected using a syringe pump and its flow was 0.25 μL/min, final volume of injected toluene in the system was 0,140 mL. The concentration of toluene was measured in constant time intervals during the dark phase and during photochemical reaction by GC (Varian 3900) coupled with the mass spectrometer (Varian Saturn 2100 T).

3. Results

3.1. Structure and morphology of films

The X–ray diffraction patterns in figure 1 illustrate that anatase with the most intense peak at $2\theta = 25.3°$ is the major crystalline phase in all cases. Additionally, the presence of amorphous silica binder in self-cleaning samples is well evident from the broad peak, while in the water- and air-cleaning samples the rutile peak is related to Degussa P25 addition.

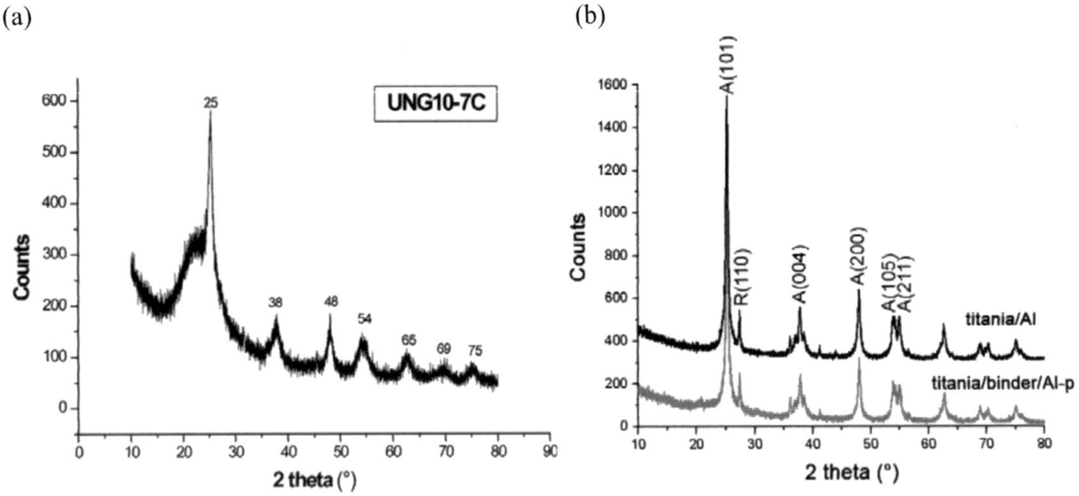

Figure 1. XRD patterns of the powder samples obtained by casting the coating solutions in petri dishes and treated in the same way as films: (a) self-cleaning samples and (b) water- and air-cleaning samples.

The SEM micrographs of all the samples are shown in figure 2. We can see that the nanostructured surface of the thin self-cleaning film (figure 2 (a) and (b)) was homogeneous and crack free, the thickness of the film deposited by dip-coating was around 200 nm. The SEM image of titania/Al without a binder component presented in figure 2 (c) shows quite big and separated aggregates, however after using the binder the surface became much more homogeneous with interconnected particles (figure 2 (d)) and the films adhered well to the substrate. The thickness of the water- and air-cleaning films was more than 10 times thicker than transparent self-cleaning films.

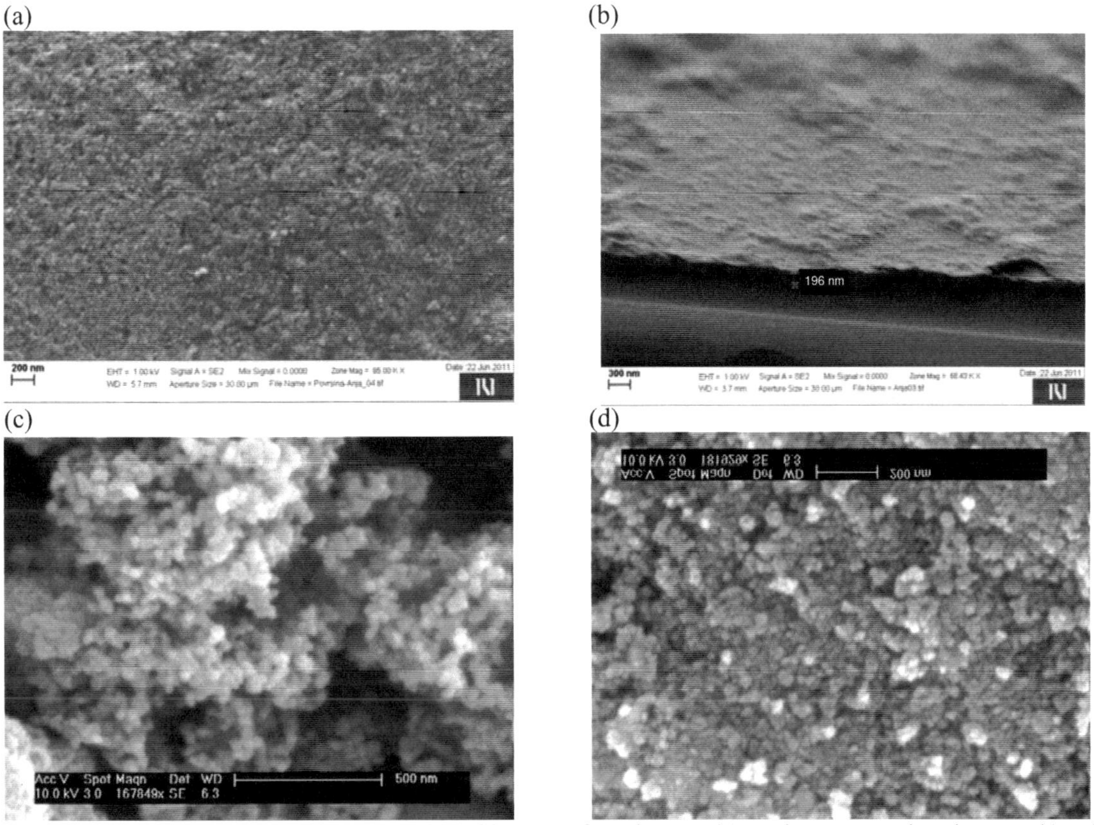

Figure 2. SEM images of the TiO_2/SiO_2 films: surface image (a) and cross section image (b) of UNG10-7Cdip on glass, surface images of titania/Al (c) and titania/binder/Al-p (d).

3.2. Self-cleaning efficiency of the thin films on glass

Figure 3 shows self-cleaning efficiencies of the samples UNG10-7Cdip, UNG10-7Cspray (fresh and old - 2 months after weathering exposure) in comparison with commercial Pilkington and St. Gobain glass. The efficiencies determined by the CA method (fig. 3 (a)) and by the fluorescence method (fig. 3 (b)) are well correlated. The water contact angles of the samples after UVA pre-irradiation and before methyl stearate deposition are low, especially in case of the most effective UNG10-7Cdip sample, which showed the strongest photoinduced superhydrophylicity effect. After methyl stearate deposition (time 0) the surface became hydrophobic and with time of irradiation the organic layer was gradually decomposed resulting in a decrease of CA. Samples UNG10-7Cspray were less effective in degrading organic layer (methyl stearate and TPA), which was mainly ascribed to lower thickness of these films. However, the efficiency was not deteriorated after ageing of the film for two months at atmospheric conditions. Note the time scale in obtaining results by the two methods – hours vs. minutes – showing an advantage of the fluorescence method over the CA method.

3.3. Photocatalytic activity of the films toward organics removal from water and air

Figure 4 shows a high activity (fast kinetics) of the titania/binder/Al-p samples toward degradation of phenol in water. With repeating photodegradation experiments using the same catalyst slides (experiments from 1 to 5), a small decrease in activity was noticed (half-life time increased from 18 to 26 min), but the coatings still remained highly photocatalytically active as the complete degradation of phenol solution was always observed within 3 hours. This was also confirmed by the TOC analysis. Total organic carbon measured after 3 hours of irradiation of phenol solution in presence of titania/binder/Al-p samples was in the range 0 – 0.5 mg/L. On the other hand, titania/Al samples were not stable under the same conditions and their activity rapidly decreased when repeating experiments.

(a) (b)

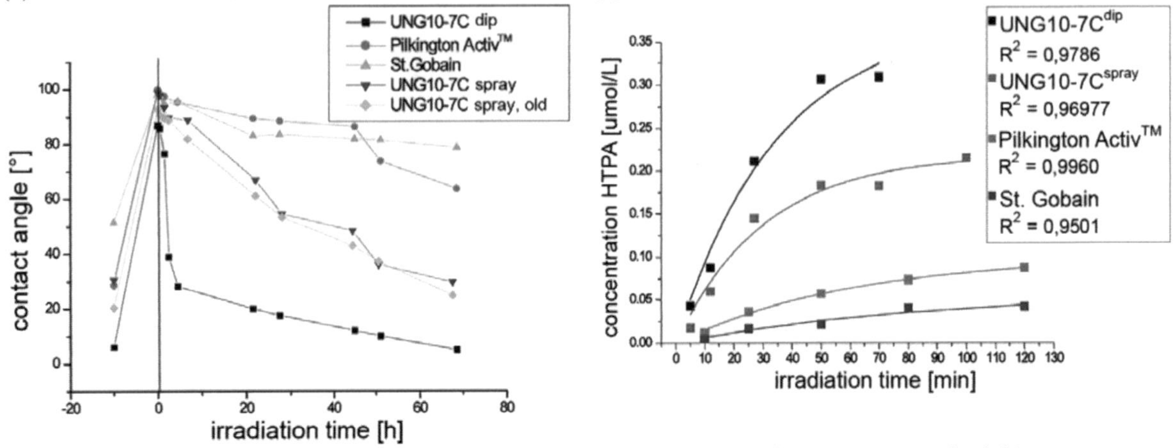

Figure 3. Self-cleaning activities determined by the CA method (a) and fluorescence method (b).

(a) (b)

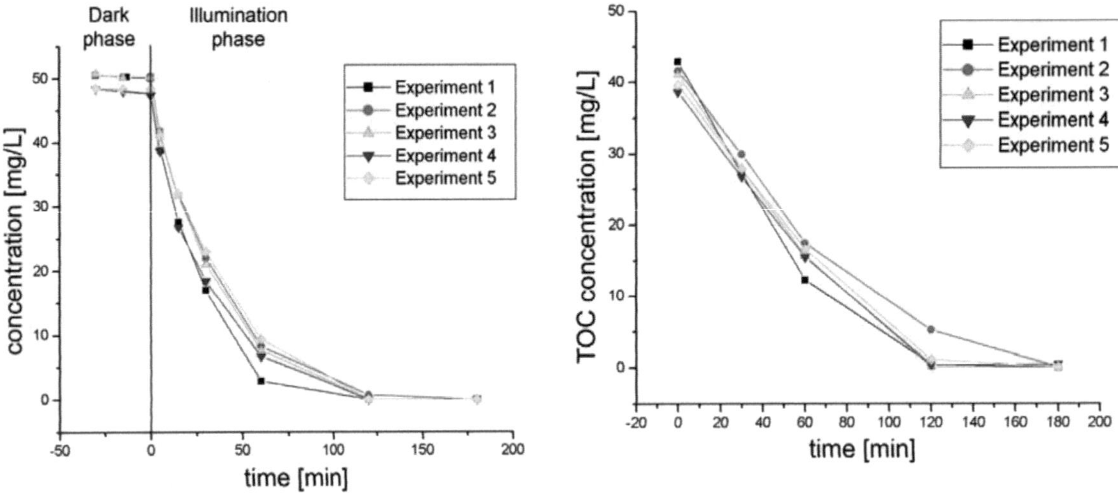

Figure 4. Photocatalytic activity of the titania/binder/Al-p samples toward degradation of phenol in water (five repetitions): HPLC analysis of phenol concentration before and during UVA irradiation (a), TOC analysis (b).

Both types of samples on Al-slides (titania/binder/Al-p and titania/Al) were also used in the gaseous photocatalytic setup toward degradation of toluene in air (figure 5). The rapid fall of toluene concentration in the dark period corresponded to its adsorption on the surface of particular sample and to its dilution in the reactor cell. As expected, the samples of bare and protected Al support (Al and Al-p) did not show any photocatalytic activities. However, again very good photocatalytic activities with fast degradation kinetics were achieved with the immobilized catalyst on Al-sheets. The binder contributed, apart from achieving better mechanical stability of the coatings (good adherence to the substrate), also to significant improvement of the dark adsorption capacity of the coatings. It is important to mention that the photocatalytic activity was not reduced with "dilution" of titania by photocatalytically inactive silica in the titania/binder/Al-p samples, having the same surface density of the material on the support as pure titania/Al samples (1 mg/cm^2).

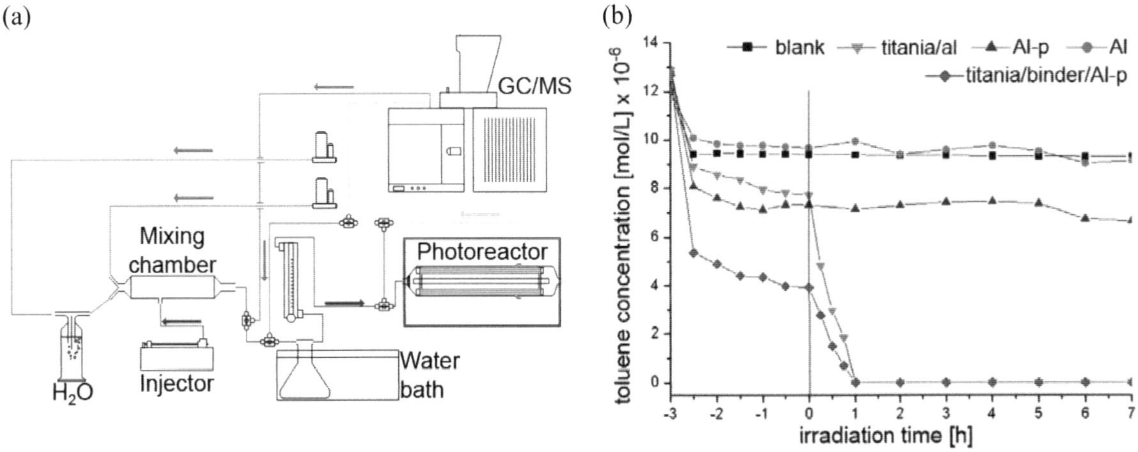

Figure 5. Schematic diagram of the set-up for gaseous photocatalysis (a), dark adsorption and photodegradation curves of gaseous toluene (b).

4. Conclusion

Solution-derived TiO_2/SiO_2 films prepared by low-temperature processing proved to be effective under near UV irradiation in removal of organic contaminants from water, air and solid deposits on their surface. An overview of our recent research on immobilized catalysts, *i.e.* thin films on glass or aluminium sheets, for environmental applications is given here (an invited lecture for EMRS 2011 fall meeting, Warsaw, Poland).

5. References

[1] Kaneko M and Okura I 2002 *Photocatalysis: Science and Technology* (Berlin: Springer)
[2] Carp O, Huisman C L and Reller A 2004 *Prog. Solid State Chem.* **32** 33
[3] Mills A and Lee S-K 2002 *J. Photochem. Photobiol. A: Chem.* **152** 233
[4] Fujishima A, Zhang X and Tryk D A 2008 *Surf. Sci. Reports* **63** 515
[5] Maver K, Lavrenčič Štangar U, Černigoj U, Gross S and Cerc Korošec R 2009 *Photochem. Photobiol. Sci.* **8** 657
[6] Černigoj U, Kete M and Lavrenčič Štangar U 2010 *Catal. Today* **151** 46
[7] Tasbihi M, Lavrenčič Štangar U, Černigoj U, Jirkovsky J, Bakardjieva S and Novak Tušar N 2011 *Catal. Today* **161** 181
[8] Černigoj U and Lavrenčič Štangar U 2009 *Preparation of TiO_2/SiO_2 sols and use thereof for deposition of self-cleaning and antifogging coatings*, international application number PCT/SI2009/000052, international publication number WO 2010/053459 A1 (Munich: European Patent Office)
[9] Soklič A 2011 *Deposition and possible influence of a self-cleaning TiO_2-SiO_2 thin film on a photovoltaic module efficiency*, diploma work (Nova Gorica: University of Nova Gorica)
[10] Šuligoj A, Černigoj U and Lavrenčič Štangar U 2010 *Preparation procedure of durable titania coatings on metal supports for photocatalytic cleaning applications*, national application number P-201000432 (Ljubljana: Urad Republike Slovenije za intelektualno lastnino)
[11] Černigoj U, Lavrenčič Štangar U and Trebše P 2007 *J. Photochem. Photobiol. A: Chem.* **188** 169
[12] Tasbihi M, Šinigoj E, Kete M and Lavrenčič Štangar U 2011 *Environ. Sci. Pollut. Res.*, submitted

IOP Conf. Series: Materials Science and Engineering **30** (2012) 012002 doi:10.1088/1757-899X/30/1/012002

Low temperature processing of alkoxide-derived PMN thin films

Tomoya Ohno[1,4], Yasunori Gotoh[2], Naonori Sakamoto[2], Naoki Wakiya[2], Takanori Kiguchi[3], Takeshi Matsuda[1] and Hisao Suzuki[2]

[1]Department of Materials Science, Kitami Institute of Technology, 165 Kouen-cho Kitami, Hokkaido, Japan 090-8507
[2]Department of Materials Science and Chemical Engineering, Shizuoka University, 3-5-1 Johoku Naka-ku Hamamatsu, Shizuoka, Japan 4328561
[3]Institute for Material Research, Tohoku University, 2-1-1 Katahira, Aoba-ku Sendai, Miyagi, Japan 980-8577

E-mail: ohno@mail.kitami-it.ac.jp

Abstract. A well-crystallized $Pb(Mg_{1/3}Nb_{2/3})O_3$ (PMN) thin film on a Si wafer was prepared using low-temperature processing with a $LaNiO_3$ (LNO) seeding layer. The PMN thin film was crystallized at 550 °C on a LNO/Si stacking structure. Additionally, we attempted to apply the compressive residual stress to PMN layers for a Curie temperature shift. Results show that the ferroelectric property remained at room temperature. The remanent polarization increased concomitantly with increasing annealing temperature.

1. Introduction

Relaxor ferroelectric materials such as $Pb(Mg_{1/3}Nb_{2/3})O_3$ (PMN) have attracted attention for application in ultrasonic and actuator devices because of their excellent electromechanical properties. Especially, (1-x)$Pb(Mg_{1/3}Nb_{2/3})O_3$-x$PbTiO_3$ (PMN-PT) solid solutions (x=0.35) have exhibited good piezoelectric properties and a high dielectric constant [1]. However, they are usually sintered severely, leading to increased vacancies and the decreased electrical properties of the resulting PMN-PT ceramics. Therefore, for these ceramics and thick and thin films in chemical solution deposition (CSD), many low-temperature sintering processes have been reported [2, 3]. However, for lower-temperature sintering and annealing, the pyrochlore phase remains in many cases, which engenders lower electrical properties. Therefore, we attempted to deposit PMN thin films at low temperatures using a reasonable seeding layer. Here the lattice constant of PMN (cubic cell) and $LaNiO_3$ (LNO) (cubic cell) are reported respectively as 0.4053 nm [4] and 0.3860 nm. Although the lattice constant of LNO is slightly smaller (about 5%) than that of PMN, LNO can be useful as a seeding layer for PMN. Our previous study demonstrated deposition of an LNO thin film with a-axis preferred orientation on a Si wafer [5]. Therefore, we selected the LNO layer with a-axis preferred orientation as a seeding layer for PMN thin film.

Furthermore, the Curie temperature of PMN was reported as approximately –11 °C [6]. The ferroelectric and piezoelectric properties disappear at room temperature. However, Choi *et al.* reported that the compressive lattice strain engenders the Curie temperature shift to higher temperature [7]. They applied the compressive stress to a barium titanate thin film using a single crystal substrate with large thermal expansion coefficient, resulting in the Curie temperature shift toward higher temperature over 400 °C. In our previous study, the residual stress in ferroelectric thin films was controlled by

[4]Tomoya Ohno: ohno@mail.kitami-it.ac.jp

designing the buffer layer structure, even on a Si wafer [8]. Here, the thermal expansion coefficients of LNO and PMN are reported respectively as 12.5 and 10.0 × 10^{-6} / °C [9]. Consequently, the PMN film on a LNO layer is expected to receive the compressive stress as thermal stress. For this study, we estimated the strain effect on electrical properties in PMN film on a commercially used Si wafer, deposited by low-temperature processing using a seeding layer.

2. Experimental Procedure

Using CSD, we deposited PMN and an LNO thin film on a Si wafer. The starting reagents for LNO precursor solution were $La(NO_3)_3$ and $Ni(CH_3COO)_2$. The experimental details have been described elsewhere [5]. Using a spin coating method, LNO layers with 200-nm thickness were deposited on a Si (100) wafer. The as-deposited LNO layer was dried at 150 °C for 10 min; then it was pre-annealed at 350 °C for 10 min to remove the residual organic compounds. Final annealing was conducted at 700 °C for 5 min using rapid thermal annealing (RTA) in an oxygen atmosphere.

The starting reagents for PMN precursor solution were $Nb(OCH_2CH_3)_5$, $Mg(OCH_2CH_3)_2$, and lead isopropoxide. Niobium ethoxide and magnesium ethoxide were dissolved into ethanol and were subsequently mixed at 78 °C for 12 hr to prepare the columbite-type precursor solution. Then the lead isopropoxide was added to the columbite-type precursor solution to prepare the PMN precursor solution. It was stirred at 78 °C for 12 hr. Finally, 2-aminoethanol was added to PMN precursor solution. The PMN precursor solution concentration was adjusted to 0.2 M. The PMN layers were deposited onto a LNO/Si substrate by spin coating. The obtained PMN layer was dried at 150 °C for 10 min and pre-annealed at 350 °C for 10 min. Final annealing was conducted at different annealing temperatures between 550–850 °C to investigate the crystallization behavior.

The crystalline phase and the lattice strain of the obtained thin film were determined using X-ray diffraction (XRD, D8 Advance; Bruker AXS GmbH). The microstructure and the film thickness were observed using an atomic force microscope (AFM, SPI3900N; SII Nano Technology Inc.) and a scanning electron microscope (SEM). The ferroelectric and dielectric properties of the resultant PMN films were measured respectively using a ferroelectric test system (Toyo Corp., FCE-PZ) and an impedance analyzer (HP-4194A; Hewlett Packard Co.).

3. Results and Discussion

3.1. XRD analysis

Figure 1 shows the XRD patterns for the obtained PMN thin films with different annealing temperatures. At 550 °C annealing, a faint amount of pyrochlore phase was identified, although the

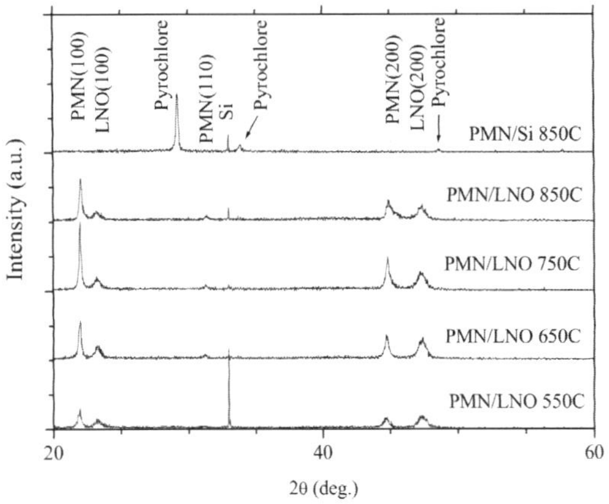

Figure 1. XRD patterns for the obtained PMN/LNO and PMN/Si stacking structures with different annealing temperatures.

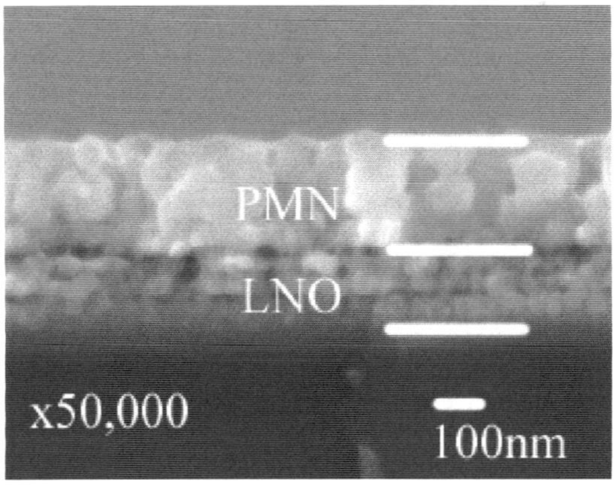

Figure 2. Cross-sectional SEM image of the PMN/LNO stacking structure.

main phase was a perovskite. In addition, the PMN films on a LNO seeding layer deposited at 650, 750 and 850 °C were identified as a completely single phase of a perovskite, although only a pyrochlore phase was confirmed in the case of the PMN/Si stacking structure, even at the high temperature of 850 °C annealing, suggesting the effect of the LNO seeding layer for low annealing of PMN thin film. Furthermore, the obtained PMN films on a LNO seeding layer exhibited a-axis preferred orientation.

Figure 2 portrays a cross section SEM image for the PMN/LNO/Si stacking structure. The respective PMN and LNO film thicknesses were determined as 250 and 200 nm. In addition, the CSD-derived LNO layer examined in our previous study included nanopores [8]. The obtained CSD-derived LNO layer also has nanopores in the LNO layers, engendering relaxation of the tensile strain from a Si wafer because of differences in the thermal expansion coefficient. In contrast, the PMN layer exhibited a dense structure.

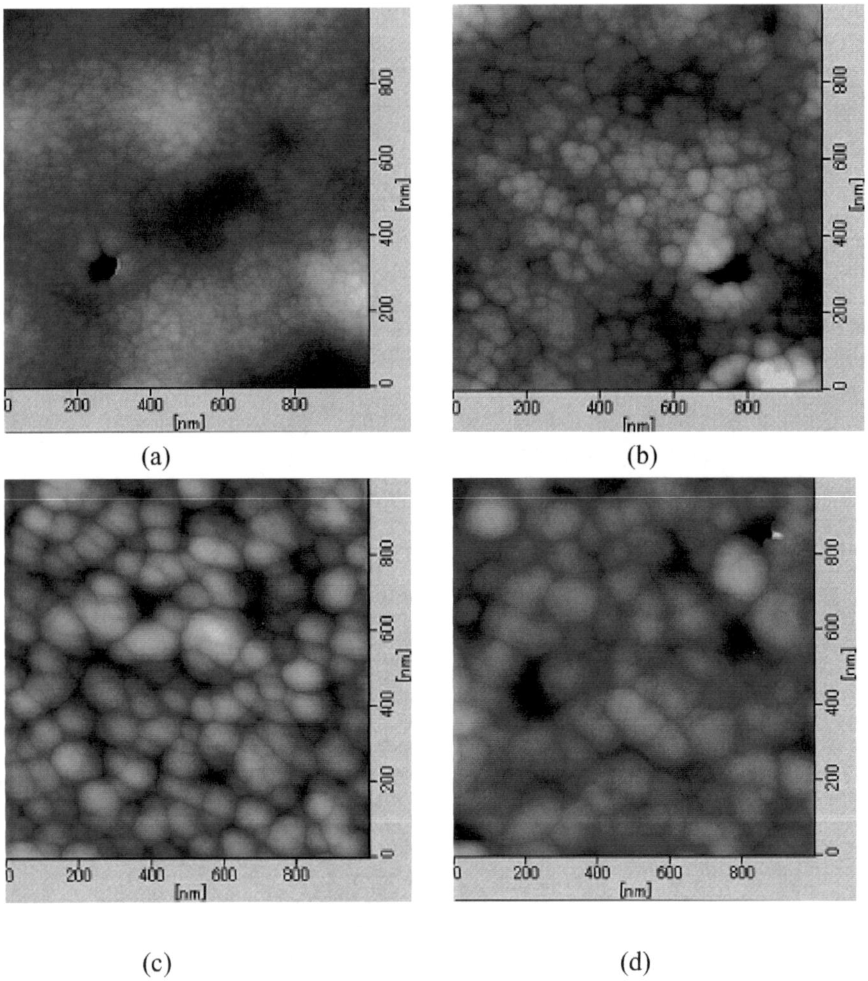

Figure 3. AFM images for the obtained PMN/LNO stacking structure with different annealing temperatures of (a) 550 °C, (b) 650 °C, (c) 750 °C, and (d) 850 °C.

3.2. Electrical Properties

Figure 3 presents AFM images for the obtained PMN films on a LNO layer. These AFM images show that the crystallite size increased concomitantly with increasing annealing temperature. The root mean square (RMS) value was approximately 5 nm, indicating a smooth surface for the obtained PMN films deposited at all annealing temperatures.

Figure 4 shows the change in the dielectric constant as a function of the measuring frequency. At 1 kHz, the dielectric constants of PMN films annealed at 550, 650, 750, and 850 °C were, respectively, 316, 690, 770, and 263. The dielectric constant increased concomitantly with increasing annealing temperature, except for the case of 850 °C annealing. In addition, Figure 5 shows P–E hysteresis loops for the obtained PMN films. The remanent polarization (Pr) increased concomitantly with increasing annealing temperature. However, at 850 °C annealing, the squareness of the hysteresis loop decreased in comparison with those of the hysteresis loops for the PMN films with different annealing temperatures.

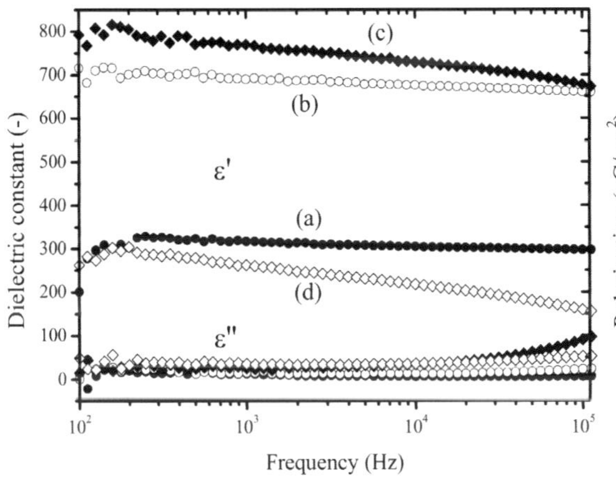

Figure 4. Dielectric constant for PMN films with different annealing temperatures as a function of measuring frequency (a) 550 °C, (b) 650 °C, (c) 750 °C, and (d) 850 °C.

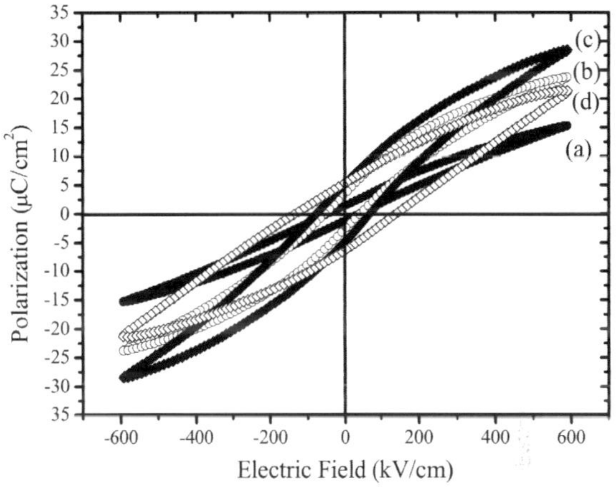

Figure 5. P-E hysteresis loop for PMN films with different annealing temperatures of (a) 550 °C, (b) 650 °C, (c) 750 °C, and (d) 850 °C.

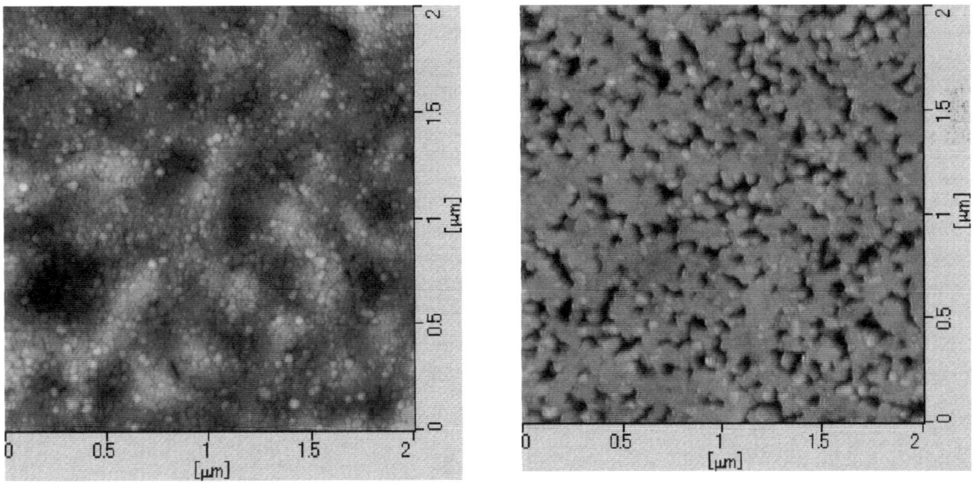

(a) (b)

Figure 6. AFM images for the LNO surface with (a) 700 °C and (b) 800 °C annealing temperatures.

To investigate the reasons for the reduced electrical properties of the PMN thin films deposited at 850 °C annealing, we confirmed the effect of the annealing temperature on the surface morphology of LNO layer. Figure 6 shows surface morphologies for the obtained LNO layers with different annealing temperatures. These images exhibit clearly that the surface morphology of LNO layers changed from dense (includes some nanopores) to a porous structure by the decomposition of the LNO crystalline phase as a seeding layer to increase the pyrochlore phase in the PMN at temperatures higher than 800 °C, probably at the interface between LNO electrode and the ferroelectric PMN layer. Actually, this phenomenon led to the increase in the resistivity of the LNO layer from 1.14×10^{-3} to 10.35×10^{-3} Ωcm by changing the annealing temperature from 700 °C to 800 °C, as well as decreasing the perovskite ratio for PMN from 97.6% at 750 °C to 94.7% at 850 °C.

3.3. Lattice Strain

The Curie temperature of PMN is reported as –11 °C. However, Figure 5 clearly illustrates that the obtained PMN films exhibited ferroelectricity at room temperature. This result indicates the compressive lattice strain was applied to the obtained PMN films from an LNO layer, and that the Curie temperature shifted to a higher temperature. Figures 7 and 8 portray the exact (200) peak position and the change in the (200) peak position with Ψ angle, respectively.

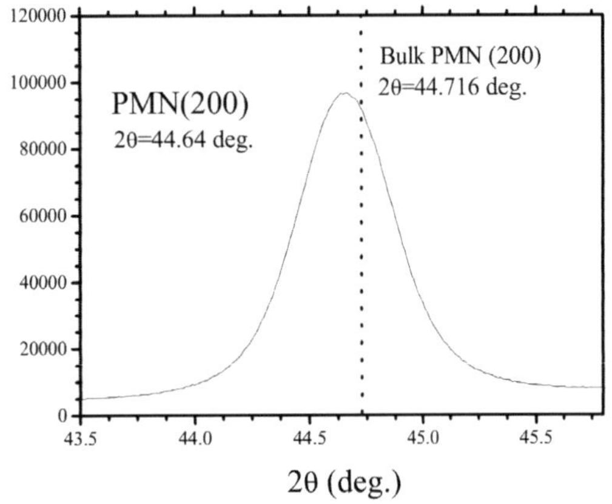

Figure 7. (200) peak position of the obtained PMN thin film deposited at 650 °C.

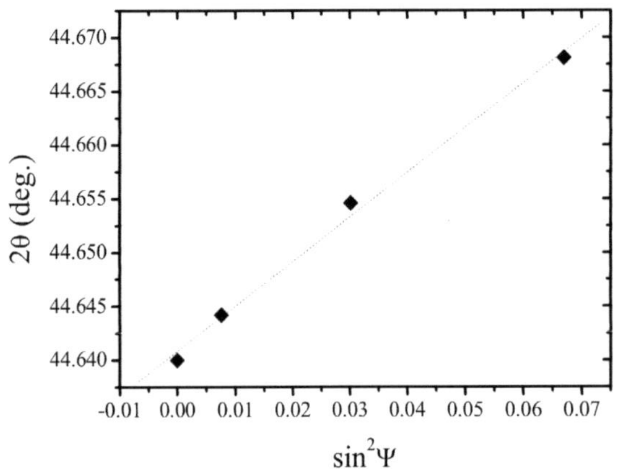

Figure 8. Change in the (200) peak position as a function of $\sin^2\Psi$.

The reported (200) peak position for the stress-free PMN is 44.716 deg [10]. However, the (200) peak position for the obtained PMN thin film was 44.64 deg. This result indicates that the out-of-plane-lattice constant increased because of the 2-D compressive stress from the LNO layer. Additionally, we estimated the shift of (200) peak position with increasing Ψ angle to calculate the residual stress in PMN thin films. Results show that the (200) peak position shifts to the higher 2θ angle with increasing ψ angle. In addition, $\partial 2\theta/\partial \sin^2\psi$ was estimated as 0.42. The Young's modulus and the Poisson's ratio for PMN are reported respectively to 61 GPa and 0.3. Therefore, we concluded that the compressive residual stress in the obtained PMN thin film with 650 °C annealing temperature was 0.18 GPa; this compressive stress was the origin of the Curie temperature shift.

4. Conclusions

The PMN thin films on a LNO seeding layer were crystallized at a lower temperature of 550 °C annealing. All obtained PMN thin films exhibited (100) preferred orientation. The electrical properties of the obtained PMN thin films increased concomitantly with increasing annealing temperatures because of the increased crystal size or crystallinity. However, with 850 °C annealing, the electrical properties decreased abruptly because of decomposition of the conductive LNO seeding layer. The compressive stress was remained in the obtained PMN thin films on a LNO layer. This compressive residual stress engenders the Curie temperature shifts to higher temperatures, resulting in ferroelectricity of the alkoxide-derived PMN thin films at room temperature.

References

[1] Uršič H, Hrovat M, Holc J, Tellier J, Drnovšek S, Guiblin N, Dkhil B and Kosec M 2010 *J. Europ. Ceram. Soc.* **30** 2081-92
[2] Costa A L, Galassi C and Roncari E 2002 *J. Europ. Ceram. Soc.* **22** 2093-2100
[3] Kumar P, Sonia R, Patel K, Prakash C and Goel T C 2008 *Mater. Chem. Phys.* **110** 7-10
[4] Lebedinskaya A R, Kupriyanov M F and Skulski R 2001 *Mater. Sci. Eng.* B**83** 119-22
[5] Miyazaki H, Goto T, Miwa Y, Ohno T, Suzuki H, Ota T and Takahashi M 2004 *J. Europ. Ceram. Soc.* **24** 1005-8
[6] Wang J, Junmin X, Dongmei W and Weibeng N 1999 *Solid State Ionics* **124** 271-9
[7] Choi K J et al. 2004 *Science* **306** 1005-9
[8] Ohno T, Malič B, Fukazawa H, Wakiya N, Suzuki H, Matsuda T and Kosec M 2009 *J. Ceram Soc. Jpn.* **117** 1089-94
[9] Bonneau P, Garnier P, Husson E and Morell A 1989 *Mater. Res. Bulletin* **24** 201-6
[10] Bonneau P, Garnier P, Calvarin G, Husson E, Gavarri J R, Hewat A W and Morell A 1991 *J. Solid State Chem.* **91** 350-61

Mesoporous sol-gel deposited SiO_2-SnO_2 nanocomposite thin films

A A Ponomareva[1, 2], V A Moshnikov[1] and G Suchaneck[2]

[1]St. Petersburg State Electrotechnical University (LETI), 197376 St. Petersburg, Russia
[2]TU Dresden Solid State Electronics Laboratory, 01062 Dresden, Germany

E-mail: ap_k@inbox.ru, Gunnar.Suchaneck@tu-dresden.de

Abstract. Nanocomposite SiO_2-SnO_2 thin films were prepared on oxidized silicon substrates using the sol-gel technique with $SnCl_2 \cdot 2H_2O$ as a tin precursor and tetraethylorthosilicate (TEOS) as a silica source. The porous structure of the films was investigated by atomic force microscope. Gas sensitivity of the samples was measured at different temperatures and for different H_2 concentrations. Advanced gas sensing properties were obtained for a silica fraction of 10 at% providing a mesoporous microstructure.

1. Introduction

Since first proposed in 1962 [1], research efforts focused on improving the gas sensitivity of oxide semiconductor-based gas sensors. The gas detection principle of semiconductor type gas sensors is based on the variation of the depletion layer at the grain boundaries in the presence of reducing or oxidizing gases [2]. Chemisorbed oxygen species (O_2^-, O^-) from the air act as electron acceptors and lead to the formation of a surface depletion layer as well as a surface barrier. Upon exposure to reducing gases, their oxidation takes place at the surface resulting in lower oxygen ion coverage. This returns electrons into the SnO_2 bulk and lowering thus the surface barrier. In consequence, a decrease of the electrical conductivity is obtained.

By means of nanosized SnO_2, the operating temperature for hydrogen gas sensors can be reduced to 100-150 °C [3] compared to conventional SnO_2 sensors typically operating in the temperature range 200–500 °C [4]. SnO_2 nanostructured materials such as nanoparticles, nanorods, nanotubes, nanobelts, nanosheets, nanocubes, mesoporous structures with well aligned pores, hollow oxide structures and hierarchical nanostructures were fabricated in order to maximize gas sensing performance by increasing the surface area to volume ratio [5]. However, for SnO_2 grains with size of about 1 to 4 nm, the grain growth starts already at temperatures of 200 to 400°C. Thus, smaller grain size results in higher sensitivity, but at the same time long-term stability of gas sensors is decreased [6]. The resistance drift of SnO_2 gas sensors becomes apparent for a grain size less than 20 nm and increases up to 40% for a grain size of 2 nm during a 30-day test [3]. On the other hand, nano-sized materials easily aggregate due to their strong and inevitable van der Waals attraction which is inversely proportional to the particle size. They irreversibly tend to form hard, secondary aggregates from densely packed agglomerates of smaller grains. This hampers interdiffusion of the sensing gas toward the sensing surface as well as the counter-diffusion of the product gases to the ambient. Thus, only the primary particles near the surface region of the agglomerates contribute to the gas sensing reaction and the inner part remains inactive [6]. Moreover, the sluggish gas diffusion through the aggregated slows the gas response speed.

Recently, one of our laboratories has applied this approach for the fabrication of SnO_2-SiO_2 composites comprising hierarchical macro-, meso- and micropores (according to IUPAC notation [7]). In this work, we investigate gas sensing properties of such sol-gel deposited nanocrystalline SnO_2-SiO_2 composites in relation to their microstructure.

2. Experimental

Nanocrystalline SnO_2-SiO_2 thin films comprising a hierarchical pore structure were fabricated using a polymeric sol-gel process on oxidized silicon and quartz glass substrates. Tin(II) chloride dihydrate $SnCl_2 \cdot 2H_2O$ (M = 225.63 g/mol) was used as a tin precursor and tetraethylorthosilicate (TEOS) $C_8H_{20}O_4Si$ (M = 208.33 g/mol) as a silica source. Solution No. 1 consisted of 10 ml (170 mmole C_2H_5OH) of 96 % ethanol as solvent, 1.07 mmole $SnCl_2 \cdot 2H_2O$ and 0.268 mmole TEOS, solution No. 2 10 ml (110 mmole) *n*-butanol as solvent, 0.683 mmole $SnCl_2 \cdot 2H_2O$ and 0.171 mmole TEOS. The solutions were aged 24 h at room temperature and spin-coated onto 3" oxidized silicon wafers and quartz glass substrates without an adhesion promoter. The films were calcined at about 550°C in air for 5 minutes by transferring the wafer into a hot furnace. Cooling down was performed in the furnace within 12 to 14 hours.

Microstructure of the deposited porous SnO_2-SiO_2 films was investigated by atomic force microscopy using an NTEGRA Therma scanning probe microscope (NT-MDT, Zelenograd, Russia). Tapping mode imaging was performed in air using silicon cantilevers (Type NSG01, NDT-MDT, Zelenograd, Russia). The specific surface area (SSA) was measured using the Brunauer, Emmett and Teller (BET) method with nitrogen adsorption using a Sorbi-N4.1 instrument (ZAO META, Novosibirsk, Russia). The UV-VIS-NIR optical properties of the films were investigated by absolute transmission measurements in a two-beam spectrophotometer Lambda 950 (PerkinElmer, USA) in the range of 300–2000 nm with steps of 2 nm. Transmission measurements of liquid samples (sols) were performed by Lambda 900 (PerkinElmer, USA) in the same range as used for films.

Sensor structures were prepared using films deposited onto oxidized silicon substrates with *n*-butanol as a solvent. Nickel-chrome and gold layer of a thickness of 10 and 35 nm, respectively, were evaporated onto the SnO_2-SiO_2 thin films. Thereby, the Ni-Cr layer serves for adhesion improvement. The electrode layers were patterned by a copper contact mask forming 0.5 and 0.4 mm wide interdigital electrodes.

Gas sensing experiments were carried out by using nitrogen (reference gas) and 1.5 vol.% hydrogen (analyte gas) in nitrogen provided by a computer controlled gas flow system. The purity of all gases was ≥ 99.999 Vol. %. The total gas flow was kept constant at 90 sccm. A gas sensor area of about 3.5 mm^2 on the silicon wafer was heated using a small platinum heater. The measurement setup is described in detail elsewhere [8].

3. Results and discussion

Optical transmission spectra of sols (solutions No.1 and No. 2) were recorded before layer fabrication (figure 1(a)) and after annealing of the corresponding films (figure 1(b)). Figure 1 demonstrates that significant differences of sol-gel system optical properties were obtained only after thermal annealing. The values of absorption edge of the samples prepared using both ethanol and *n*-butanol solvents were 3.4 eV and 4.0 eV, respectively. In the *n*-butanol case the band gap is larger than the bulk value of 3.6 eV due to the contribution of quantum size effect of the present SiO_2-SnO_2 thin film [9, 10]. Moreover, the *n*-butanol solvent exhibits a higher viscosity (2.578 mPa·s at 298K [11]) compared to ethanol (1.1 mPa·s at 298K [12]). Therefore, motion and aggregation of the tin-containing clusters are slower resulting in a smaller grain size of *n*-butanol derived films.

Figure 2 shows the AFM-images of the samples. Practically no porosity was observed in films prepared without TEOS addition (fig. 2(a)). The 10 at% SiO_2-90 at% SnO_2 sample exhibits the largest porosity with a three-dimensional film structure (fig. 2(b)). The sample with 20 at% silica fraction had circular pores of different sizes (fig. 2(c)). The SSA of this film amounted to 87.21 m^2/g yielding an average pore size of about 10 nm. SSA and average pore size are in good agreement with the values of similar films investigated in [13]. On the other hand, it has to be mentioned that these nanopores known to enhance gas sensitivity [14] cannot be imaged by our AFM device. A more detailed analysis of film microstructure based on fractal analysis of surface topography will be given in a forthcoming paper.

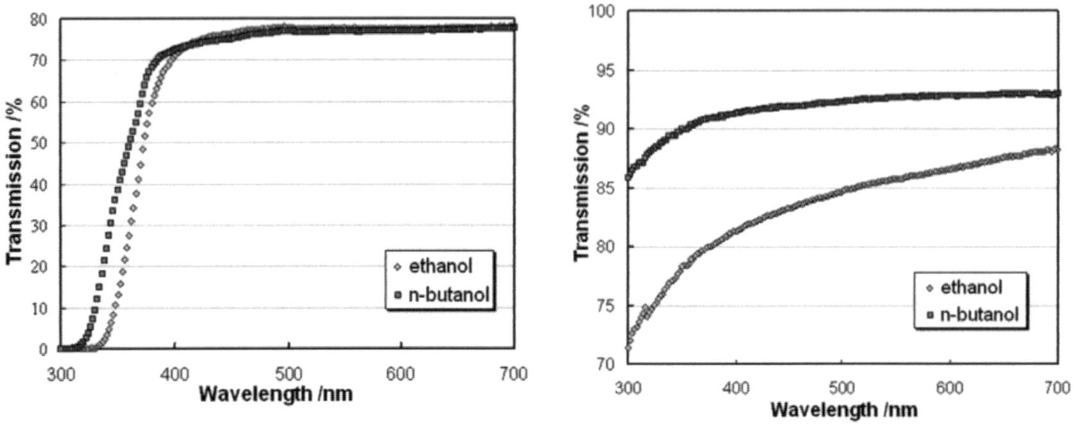

(a) (b)

Figure 1. UV-VIS optical transmission of gels prepared using different solvents (a) and of corresponding sol-gel prepared films on quartz glass substrates (b).

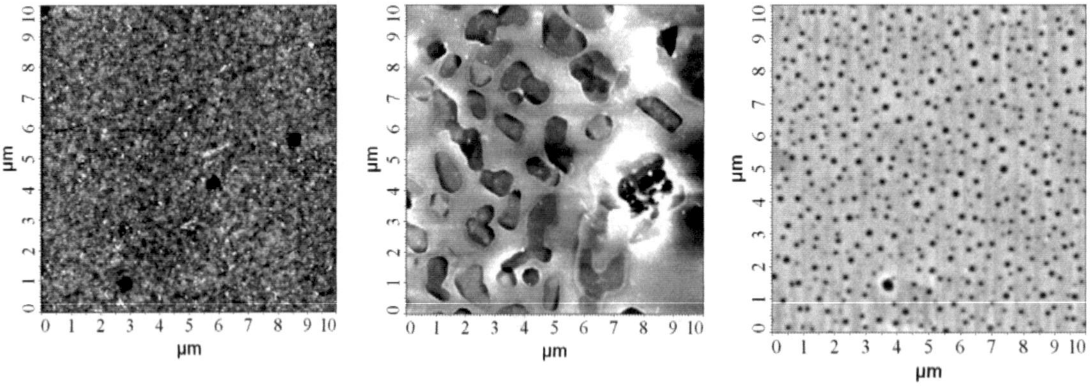

(a) (b) (c)

Figure 2. AFM-images of sol-gel layers:
(a) – 100 at% SnO_2; (b) – 10 at% SiO_2 – 90 at% SnO_2; (c) – 20 at% SiO_2 – 80 at% SnO_2.

The results of hydrogen gas sensitivity measurements at 200°C are summarized in table 1. The highest sensitivity was achieved by the sensor with 10 at% silica fraction possessing the highest porosity.

Table 1. Sensitivity (S) of tin oxide thin film sensors versus hydrogen concentration.

Composition	$S=\Delta R/R_{air}$			
	4500ppm	7500ppm	10500ppm	15000ppm
100 at% SnO_2	0,38	0,42	0,45	0,48
10 at% SiO_2-90 at% SnO_2	0,85	0,87	0,9	0,93
20 at% SiO_2-80 at% SnO_2	0,76	0,79	0,83	0,86

Figure 3(a) shows the response curve of the corresponding sensor and figure 3(b) the response time in dependence on hydrogen concentration and silica fraction. The response time exhibits the typical dependence on the gas concentration, i.e., an increase of the response time with decreasing of gas concentration [15, 16]. As in the case of hydrogen sensitivity, the 10 at%SiO_2-90 at%SnO_2 sample provides the best properties, i.e., the fastest time response.

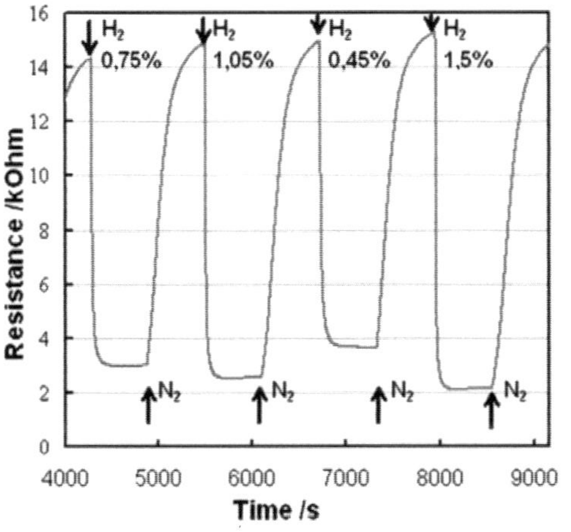

(a)

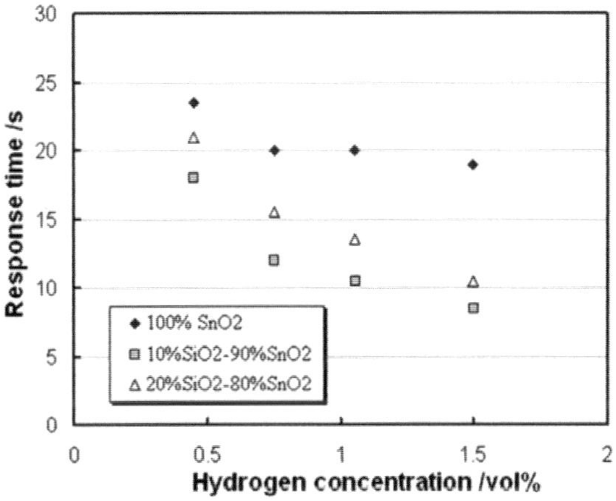

(b)

Figure 3. Response curve as obtained for the 20 at% SiO_2-80 at% SnO_2 sample (a) and dependence of response times on gas concentration (b) at a temperature of 200°C.

The hydrogen sensitivity of optimized sol-gel deposited films of this work (S = 0.85-0.93 cp. table 2) is sufficiently higher than that of commercial sol-gel fabricated gas sensors comprising Pt as catalyst (S = 0.52 at 200°C, 100 ppm H_2 [2], S = 0.32 at 200°C, 4000 ppm H_2 [17]). They exceed hydrogen sensitivity of thick films manufactured using pastes comprising nanometer sized grains of sol-gel fabricated tin oxide powders [18-20]. Comparable hydrogen sensitivity values were obtained in thin sol-gel films consisting of grains < 10 nm [21]. We attribute the somewhat higher hydrogen sensitivities (up to 0.98) reported for sol-gel films in [22-24] to an oxygen containing carrier gas (dry or synthetic air).

In order to test long-term stability of the hydrogen sensors fabricated in this work, seven hundred measuring cycles (cp. fig. 3) were carried out. No degradation of sensitivity was observed. Longer term investigations are in progress now.

4. Conclusions

Sol-gel deposition of oxide layers exhibiting advanced gas sensing properties requires an optimization of the porous film structure. In an optimized microstructure, nanopores provide high gas sensitivity due to an increased specific surface area and mesopores provide an efficient analyte transport decreasing sensor response and recovery times. SiO_2-SnO_2 nanocomposites comprising hierarchical macro-, meso- and micropores were fabricated using TEOS as a silicon source. TEOS is known to favour the formation of three-dimensional network of linked oxide particles in the gel leading to a proper pore structure in the annealed 10 at%SiO_2-90 at%SnO_2 composites.

Acknowledgements

This work was supported by contracts 16.740.11.0211, P399 and P2279 of the Federal Target Program "Scientific and Scientific-Pedagogical Personnel of Innovative Russia for the period 2009-2013", and by the Erasmus Mundus External Co-operation Window Programme of the European Union (A.P.).

References

[1] Seiyama T, Kato A, Fujiishi K and Nagatani M 1962 *Anal. Chem.* **34** 1502
[2] Gong J, Fei W, Xia Z, Chen Q, Seal S and Chow L C 2003 *Proc. IEEE* **1** 124
[3] Lu F, LiuY, Dong M and Wang X 2000 *Sensors and Actuators* B **66** 225
[4] Malyshev V V and Pislyakov A V 2008 *Sensors and Actuators* B **134** 913
[5] Lee J-H 2009 *Sensors and Actuators* B **140** 319

[6] Korotcenkov G 2005 *Sensors and Actuators* B **170** 209
[7] Rouquerol J, Avnir D, Fairbridge C W, Everett D H, Haynes J M, Pernicone N, Ramsay J D F, Sing K S W and Unger K K 1994 *Pure & Appl. Chem.* **66** 1739
[8] Delan A, Karuppasamy A and Schultheiß E 2009 *Plasma Processes and Polymers* **6** S731
[9] Gu F, Wang S F, Lu M K, Cheng X F, Liu S W, Zhou G J, Xu D and Yuan D R 2004 *J. Cryst. Growth* **262** 182
[10] Gaidi M, Hajjaji A, Smirani R, Bessais B and El Khakani M A 2010 *J. Appl. Phys.* **108** 063537
[11] Maharolkar A P, Sudke Y, Kamble S and Tidar A 2010 *International Journal of Chemistry* **2** 250
[12] Pires R M, Costa H F, Ferreira A G M and Fonseca I M A 2007 *J. Chem. Eng. Data* **52** 1240
[13] Gracheva I E, Moshnikov V A, Karpova S S and Maraeva E V 2011 *J. Phys: Conf. Ser.* **291** 012017
[14] Yang J, Hidajat K and Kawi S 2008 *Materials Letters* **62** 1441
[15] Liewhiran C, Tamaekong N, Wisitsoraat A and Phanichphant S 2009 *Sensors* **9** 8996
[16] Cukrov L M, McCormick P G, Galatsis K and Wlodarski W 2001 *Sensors and Actuators* B **77** 491
[17] Gong J, Chen Q, Fei W, Seal S 2004 *Sensors and Actuators* B **102** 117
[18] Wang Y-D, Ma C- L , Wu X-H, Sun X-D and Li H-D 2002 *Sensors and Actuators* B **85** 270
[19] Pan Q, Xu J, Dong X and Zhang J 2000 *Sensors and Actuators* B **66** 237
[20] Kim H-R, Choi K-I, Lee J-H and Akbar S A 2009 *Sensors and Actuators* B **136** 138
[21] Hammond J W and Liu C-C 2001 *Sensors and Actuators* B **81** 25
[22] Baik N S, Sakai G, Miura N and Yamazoe N 2000 *Sensors and Actuators* B **63** 74
[23] Sakai G, Baik N S, Miura N and Yamazoe N 2001 *Sensors and Actuators* B **77** 116
[24] Aroutiounian V 2007 *International Journal of Hydrogen Energy* **32** 1145

IOP Conf. Series: Materials Science and Engineering **30** (2012) 012004 doi:10.1088/1757-899X/30/1/012004

Bi-Sr-Co-O thermoelectrics prepared by sol-gel methods with modified gel decomposition

K Rubešová[1], T Hlásek[1], V Jakeš[1], D Sedmidubský[1] and J Hejtmánek[2]

[1]Department of Inorganic Chemistry, Institute of Chemical Technology Prague, Technická 5, Prague 6, 166 28, Czech Republic
[2] Institute of Physics, Czech Academy of Science, Cukrovarnická 10, Prague 6, 162 00, Czech Republic

E-mail: rubesovk@vscht.cz

Abstract. We prepared misfit $Bi_2Sr_2Co_{1.8}O_x$ (Bi-222) phase as a member of a Bi-Sr-Co-O family. Two water based sol-gel methods were chosen with regard to the presence of a strongly hydrolysing Bi^{3+} ion - chelating route combining EDTA (ethylenediaminetetraacetic acid) and TEA (triethanolamine) and, secondly, water soluble polymer method using PEI (polyethylenimine). We focused on the influence of gel decomposition process on the grain size of precursor and, consequently, on the bulk density of the final samples. We tested decomposition in N_2 atmosphere followed by a treatment in pure oxygen. The precursors decomposed in "N_2/O_2" regime were mainly composed of Bi_2O_3, $SrCO_3$ and cobalt oxides with the grain size of approximately 50-100 nm. The powders arising from gel decomposition in air contained the desired Bi-Sr-Co-O oxide as the major phase independently of the chosen sol-gel method. The final sintered samples were almost single-phase with traces of the other pseudoternary phase $Bi_2Sr_2CoO_x$ independently of the decomposition atmosphere. For comparison, samples were also prepared by solid state reaction. The sol-gel prepared samples were always of higher bulk density with larger grains, moreover partly microstructurally ordered. These facts were also reflected in transport thermoelectric measurements.

1. Introduction

As the first oxide thermoelectric material, layered Na_xCoO_2 system with the high valence of cobalt was investigated [1]. Related cobalt oxides (i.e. $Ca_3Co_{3.95}O_x$, $Bi_2Sr_2Co_{1.8}O_x$) with the general formula $[M_nA_2O_{n+2}]_x[CoO_2]_y$ have been developed so far [2, 3]. The structure generally consists of two crystal subsystems which are mutually incommensurate with respect of at least one lattice parameter. The first one with the rock salt structure is composed of square-planar MO layers (M = Co, Bi, Pb, Tl...) surrounded by AO planes also with square symmetry (A = Ca, Sr, Ba...). The CoO_2 subsystem reveals hexagonal symmetry (CdI_2 crystal type) and is built from edge sharing CoO_6 octahedra [4]. In addition, another Bi-Sr-Co oxide was prepared: $Bi_2Sr_2CoO_x$ [5]. This phase has an orthorhombic modulated layered structure with a stacking distorted perovskite type block (Sr/O-Co/O-Sr/O) and a rock salt type block ($Bi/O_{1+\delta}$). However, this phase is still out of interest concerning its potential thermoelectric properties.

To reach high thermoelectric performance, high figure of merit $ZT \sim S^2T/\rho\kappa$ is necessary (S - thermopower; T - temperature; ρ - electrical resistivity; κ - thermal conductivity). Therefore the thermoelectric material should meet some microstructural criteria, mainly bulk density. Sol-gel methods can be successfully used for preparation of oxide precursor powders that possess high chemical homogeneity, small grain size, and high reactivity.

In this work, phase $Bi_2Sr_2Co_{1.8}O_x$ was prepared using a sol-gel process. According to our previous experience with Bi-based superconducting cuprates [6], we applied two different sol-gel processes: an improved chelate method using the combination of ethylenediaminetetraacetic acid (EDTA) and triethanolamine (TEA) and, secondly, a water soluble polymer method applying polyethylenimine (PEI) as a N-donor chelating agent suitable for Bi^{3+} cation. Gels prepared by chosen sol-gel methods were decomposed either in air or inert atmosphere followed by a treatment in pure oxygen. The phase composition and grain size of the obtained powders were compared. Further the influence of decomposition atmosphere on the final samples microstructure was evaluated. Thermoelectric transport characteristics were also measured.

2. Experimental procedure

Samples with the following stoichiometry were prepared: $Bi_2Sr_2Co_{1.8}O_x$ (further called Bi-222). Two different sol-gel methods and also solid state reaction were used; each method is described separately:

i) *chelating EDTA/TEA sol-gel route:* $Bi(CH_3COO)_3$ was dissolved in acetic acid; then EDTA was added and pH was increased using ammonia until EDTA was fully dissolved; next, TEA was added to this solution. The second solution was prepared dissolving $Sr(CH_3COO)_2 \cdot 1/2H_2O$ and $Co(CH_3COO)_2 \cdot 4H_2O$ in water; EDTA was added and also deprotonated using ammonia; TEA was inserted at the end, as well. The molar ratio of chelating agent 1 : 1 : 1 (metals : EDTA : TEA) was chosen. Both prepared solutions were mixed together and heated at 80 °C. After water evaporation, the temperature was increased to 100 °C to promote esterification process and gelation [6].

ii) *water soluble polymer method:* the chosen amount of polymer was 2 monomer units per one metal cation. 50% water solution of PEI with average molecular weight of 2000 g/mol was used. $Bi(CH_3COO)_3$ was dissolved in acetic acid; then PEI solution was added; no pH adjustment was required. $Sr(CH_3COO)_2 \cdot 1/2H_2O$ and $Co(CH_3COO)_2 \cdot 4H_2O$ were added directly to the Bi-solution. The mixture was evaporated at 80 °C and then temperature was increased to 100 °C to reach gelation.

The gels obtained from both sol-gel methods were further processed as follows: after the first decomposition step at 250 °C in air, the gels were heat treated either in air (500 °C/2 h and 800 °C/2 h) or using "N_2/O_2" regime (800 °C/2 h/N_2 and 400 °C/1 h/ O_2).

Powders prepared using a sol-gel process were pressed into pellets that were finally sintered in air. The temperature and time of final sintering was 870 °C and 96 h.

iii) *solid state reaction:* Bi_2O_3, $SrCO_3$ and Co_3O_4 were homogenized in an agate mortar and then the mixture was calcined at 800 °C for 12 h. After homogenization, the calcined powder was pressed into pellets that were finally sintered in air in the same way as in the sol-gel routes.

X-ray powder diffraction data were collected at room temperature with Bruker AXS D8 θ-θ powder diffractometer with parafocusing Bragg-Brentano geometry using CoK_α radiation (λ = 1.79021 Å, U = 34kV, I = 20mA). SEM was carried out using TESCAN Vega3 equipped with EDS spectrometer. WTW Multilab 545 pH meter equipped with a glass electrode was used for pH control. Transport properties (electrical resistivity, thermal conductivity and Seebeck coefficient) were measured using four-probe method (self-designed equipment). The measurement was carried out in a vacuum cell integrated into a closed-cycle helium cryostat. The bulk density was determined by volume and weight measurement. The theoretical value of bulk density 6.81 g/cm^3 was taken for the Bi-222 phase.

3. Results and discussion

The samples prepared from gels decomposed in air (independently of the chosen sol-gel method) contained the prepared phase Bi-222 as early as after heat treatment at 500°C/2 h and 800°C/2 h. It confirms high reactivity and stoichiometric homogeneity of sol-gel precursors. Figure 1(a) shows the XRD diffraction pattern of a precursor synthesized from EDTA/TEA with gels decomposed in air.

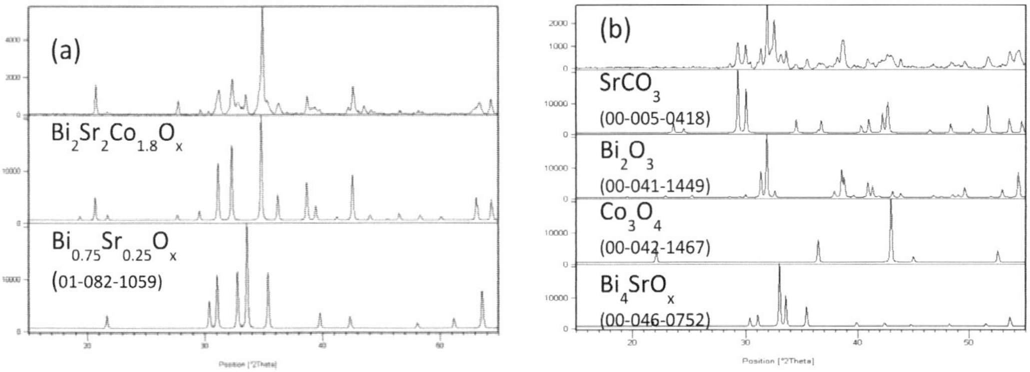

Figure 1. XRD patterns of the Bi-222 EDTA/TEA sol-gel precursor powders after gel decomposition (a) in air, (b) in inert atmosphere followed by the treatment in pure oxygen. The reference XRD pattern of Bi-222 phase was calculated using JANA 2006 program for the Rietveld refinement of modulated structures. The structural data of the phase were taken from [4].

According to our experience, $Bi_{0.75}Sr_{0.25}O_x$ phase was detected as the most frequent secondary phase present in this system at the stage of precursor. Figure 2(a) depicts the SEM image of the same powder. The grain size seems to be around 1 μm and grains are partly sintered.

On the other hand, the gels decomposed under "N_2/O_2" regime do not contain the Bi-222 final phase. These precursor powders were mainly composed of Bi_2O_3, $SrCO_3$ and cobalt oxides with the presence of various binary oxides of Bi, Sr and Co (figure 1b). Figure 2(b) shows SEM micrograph of the same powder with grain size of approximately 50-100 nm (for higher clarity, the same magnification is used and only agglomerates of grains are visible).

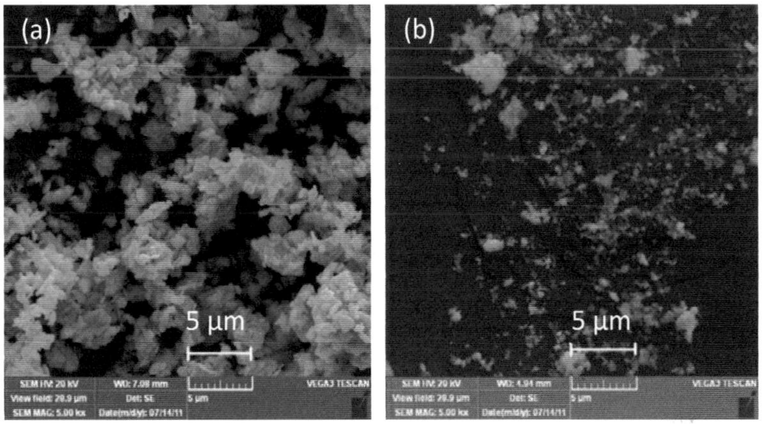

Figure 2. SEM micrographs of precursor powders obtained from EDTA/TEA gels decomposed (a) in air, (b) in inert atmosphere followed by treatment in pure oxygen. Magnification 5000x.

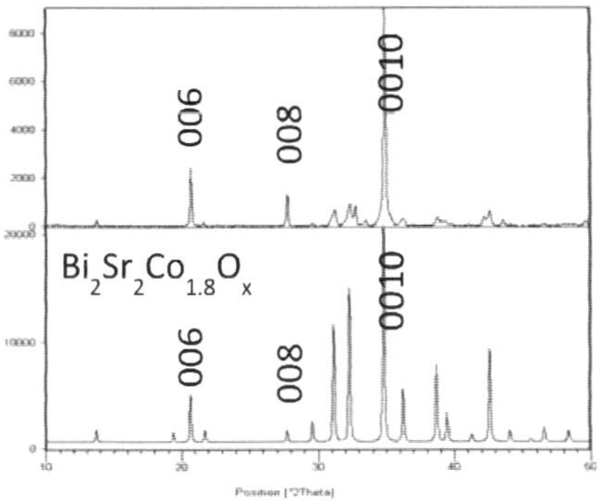

Figure 3. XRD pattern of EDTA/TEA sol-gel prepared final sample.

We tested different temperatures and times of this two-step heat treatment (not presented) and, with the aid of Organic Elemental Analysis and DTA/TG measurement, we have chosen the temperature of the oxygen treatment as low as possible to prevent agglomeration and, simultaneously, to remove fully the carbon matter – 400 °C/2 h.

With regards to the phase composition of the sintered samples, there were no differences between samples prepared in different gel decomposition atmosphere. Figure 3 shows the example of the Bi-222 final sample with enhanced 00l reflections. It is remarkable that the Bi-222 phase contained $Bi_2Sr_2CoO_x$ phase as a trace phase. These two phases are apparently in equilibrium. Also $Bi_{0.75}Sr_{0.25}O_x$ phase was found as a trace phase almost in all samples.

The final sol-gel samples with gels decomposed in air were always more dense than the samples prepared by solid state reaction – e.g. the bulk density of Bi-222 PEI sol-gel sample and of solid state sample reached 93% and 87% of the theoretical value, respectively (see figure 4(a) and (c)). The bulk density of samples with gels decomposed in "N_2/O_2" regime was lower than those prepared by solid state reaction (figure 4 (b)). At this point, the advantage of small highly reactive grains of precursor powder is lost because of the presence of strontium carbonate that, when decomposed at sintering, disrupts the microstructure of sintered pellets. Therefore we decided to incorporate an intermediate grinding and pressing step into the sintering procedure. Afterwards, the prepared samples reached similar bulk density as sol-gel samples with gels decomposed in air (a SEM picture not presented).

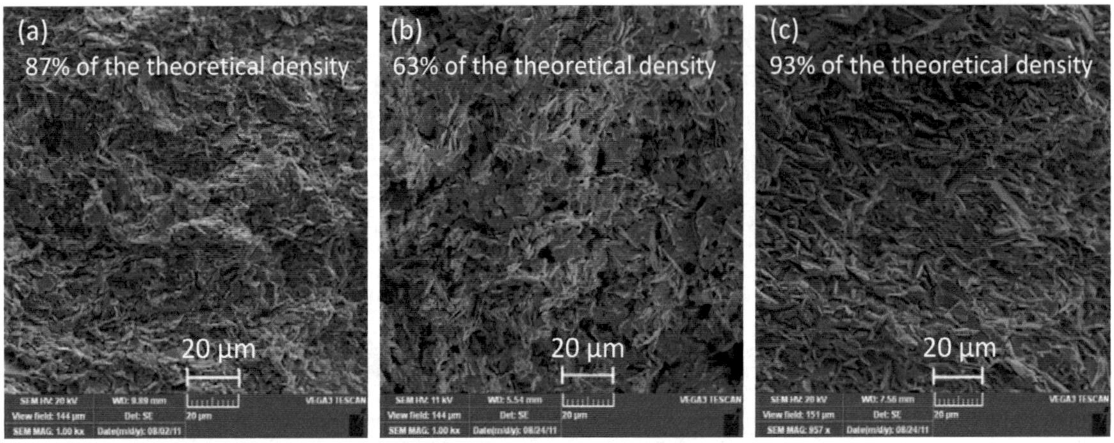

Figure 4. SEM micrographs of the sintered samples prepared either by solid state reaction-(a) or by PEI sol-gel process-(b),(c). The sample (b) was prepared from gel decomposed in"N_2/O_2", whereas the sample (c) was sintered from gel decomposed in air. Magnification 1000x is the same for all pictures.

Comparing the basic thermoelectrical parameters, sol-gel and solid state prepared samples differed neither in the values of Seebeck coefficient (figure 5(a)) nor in the dependence of thermal conductivity (not presented). The main difference was in the values of electrical resistivity (figure 5(b)) that caused also difference in figure of merit ZT ~$S^2T/\rho\kappa$ (figure 6). The results are in agreement with differences in bulk density.

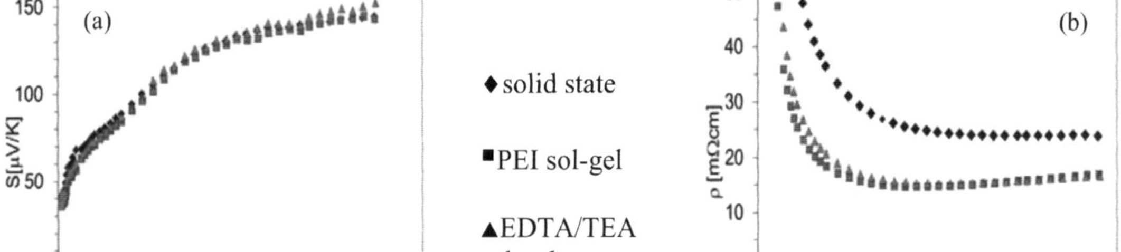

Figure 5. The temperature dependence of (a) Seebeck coefficient and (b) electrical resistivity. Both sol-gel samples were formed from the gels decomposed in air.

Thermoelectric characteristics did not differ significantly for samples prepared by different sol-gel routes. Also sol-gel samples with gels decomposed either in air or in "N_2/O_2" regime did not vary much. The figure of merit does not reach the values of thermoelectric materials prepared by methods with enhanced grain growth (flux flow sintering), however, the small grain size, high homogeneity and high reactivity of sol-gel precursors can be well utilized in techniques such as isostatic pressing etc.

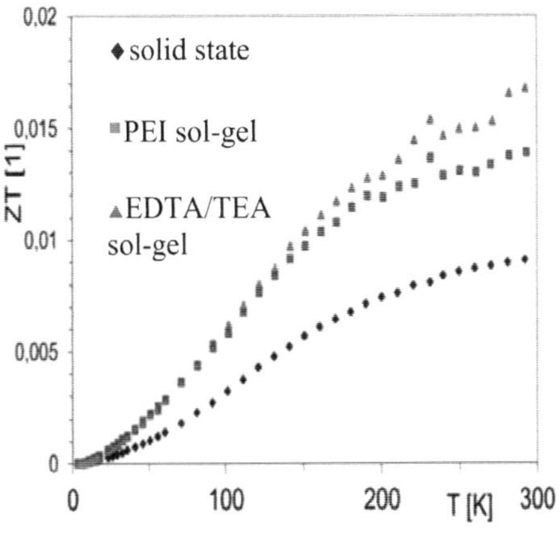

Figure 6. The comparison of the figure of merit for Bi-222 samples prepared by solid state procedure and by sol-gel methods.

4. Conclusions

The misfit $Bi_2Sr_2Co_{1.8}O_x$ phase was prepared using two different sol-gel methods – EDTA/TEA chelating method and PEI polymer route. Gels were decomposed either in air or in inert atmosphere followed by heat treatment in pure oxygen. Precursor powders prepared by the second mentioned procedure had much smaller grain size independently of the chosen sol-gel method. The precursor powders differed also in phase composition. After the gel decomposition in air (500 °C/2 h, 800 °C/2 h), the powders were mainly composed of the desired phase. This fact confirms high reactivity and homogeneity of sol-gel prepared precursors. On the other hand, the powders obtained from "N_2/O_2" regime contained mainly Bi_2O_3, $SrCO_3$ and cobalt oxides with the presence of several binary oxides. The presence of $SrCO_3$ caused that the microstructure of the final samples was destroyed after sintering. The problem was solved by intermediate grinding in the middle of sintering. The final ceramics differed in bulk density and microstructure. Bulk density and also measured thermoelectric transport characteristics were always enhanced when the sol-gel methods were used. Parameters of the sol-gel samples prepared from the gels decomposed either in air or in "N_2/O_2" regime (with intermediate grinding) were comparable. The figure of merit of the sol-gel samples was always higher in comparison to the solid state reaction.

Acknowledgement
This work was financially supported by the Czech Science Foundation, the Project No. 203/09/1036.

References

[1] Terasaki I, Sasago Y and Uchinokura K 1997 *Phys. Rev.* B **56** 12685-7
[2] Funahashi R, Matsubara I and Sodeoka S 2000 *Appl. Phys. Lett.* **76** 2385-7
[3] Li S, Funahashi R, Matsubara I, Ueno K and Yamada H 1999 *J. Mater. Chem.* **9** 1659-60
[4] Leligny H, Grebille D, Perez O, Masset A C, Hervieu M and Raveau B 2000 *Acta Crystallogr. Sect. B-Struct. Sci.* **56** 173-82
[5] Tarascon J M et al. 1989 *Phys. Rev. B* **39** 11587-98
[6] Rubesova K, Jakes V, Hlasek T, Vasek P and Matejka P 2011 *Journal of Chemistry and Physics of Solids* in press, doi:10.1016/j.jpcs.2011.11.022

IOP Conf. Series: Materials Science and Engineering **30** (2012) 012005 doi:10.1088/1757-899X/30/1/012005

Influence of substrate bias voltage on the physical, electrical and dielectric properties of RF magnetron sputtered TiO_2 films

P Kondaiah[1], M Chandra Sekhar[1], S V Jagadeesh Chandra[2], R Martins[2] S Uthanna[1] and E. Elangovan[2]

[1]Department of Physics, Sri Venkateswara University, Tirupati – 517 502, India

[2]CENIMAT/I3N, Departamento de Ciência dos Materiais, Faculdade de Ciências e Tecnologia, FCT, Universidade Nova de Lisboa and CEMOP/UNINOVA, 2829-516 Caparica, Portugal

E-mail: uthanna@rediffmail.com

Abstract. Titanium oxide (TiO_2) finds potential applications in various fields such as solar cells, optical coatings due to its high refractive index and it also has been widely used in memory devices owing to its high dielectric constant. TiO_2 films have been deposited on p-type silicon (100) substrates by RF magnetron sputtering technique. Thickness, structure and surface morphology of the films were analyzed by using α-step profilometer, Raman spectroscopy and atomic force microscope respectively. Thin film capacitors of the type $Al/TiO_2/Si$ were fabricated by evaporation of Aluminium on to the TiO_2 films. The current – voltage and capacitance - voltage characteristics were carried out to understand the electrical conduction and dielectric properties of the deposited films with a stack of $Al/TiO_2/Si$. The leakage current density was decreased and capacitance was increased with increase of substrate bias voltage.

1. Introduction

Continuous down scaling of complementary metal oxide semiconductor (MOS) devices consisting of silicon based gate dielectrics led to the high leakage currents [1]. To replace the conventional SiO_2, high dielectric constant (high-K) gate dielectrics such as Ta_2O_5, ZrO_2 HfO_2, Al_2O_3 and TiO_2 have been extensively investigated [2-4]. Among these high-K materials, TiO_2 is one of the most promising candidates due to its high dielectric constant and low leakage currents [5]. The dielectric properties of the films were mainly influenced by the deposition conditions in magnetron sputtering method. In this investigation, Titanium dioxide (TiO_2) thin films were deposited by using RF magnetron sputtering at different substrate bias voltages. The substrate bias voltage effectively influences the structural and electrical properties of TiO_2 capacitors.

2. Experimental

The TiO_2 films were deposited on silicon substrates by sputtering of titanium target of 50 mm diameter with RF power of 200 W, an oxygen partial pressure of $4x10^{-2}$ Pa and sputter pressure of 5 Pa. Then Al top contacts (300 μm dia.) were evaporated on to the TiO_2 films. The thickness of the films was measured by using Veeco Dektak α-step profilometer (Model 150), the surface morphology of the films were analyzed by atomic force microscopy (Digital instruments AFM Nanoscope III) and micro structural analysis was carried out using Raman spectrometer (Horiba Jobin Yvon Lab RAM model HR800UV). The Capacitance–Voltage (C-V) and the Current–Voltage (I-V) behavior of Metal–Oxide–Semiconductor (MOS) structures of Al/TiO_2/Si were measured at room temperature by using MIOKI (model 3532-50) LCR meter and Hewlett Packard (model hp 4140B) pA meter respectively.

3. Results and discussion

Micro Raman spectra of the TiO_2 films deposited at different substrate bias voltages are shown in fig. 1. The un-biased films have no characteristic peaks of TiO_2 phases which enable the amorphous nature of the films. For the films deposited at negative substrate bias voltages -100 and -150 V, three peaks at 144, 399 and 640 cm^{-1} were observed and are assigned to the anatase phase (E_g modes) of TiO_2 [6]. As the substrate bias voltage increased from -100 to -150 V the intensity of the peaks was found to be increased indicating that the crystallanity was increased. Karunagaran et. al [7] observed the improvement of crystallinity with increase of annealing temperature from 673 to 773 K respectively. Typical high frequency C-V curves of samples deposited at different substrate bias voltage are presented in fig. 2.

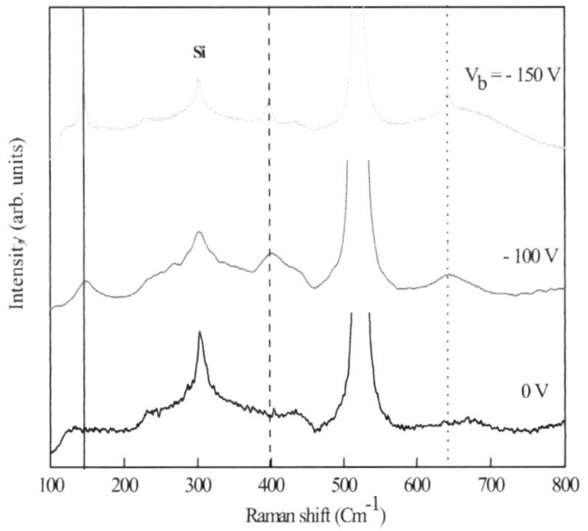

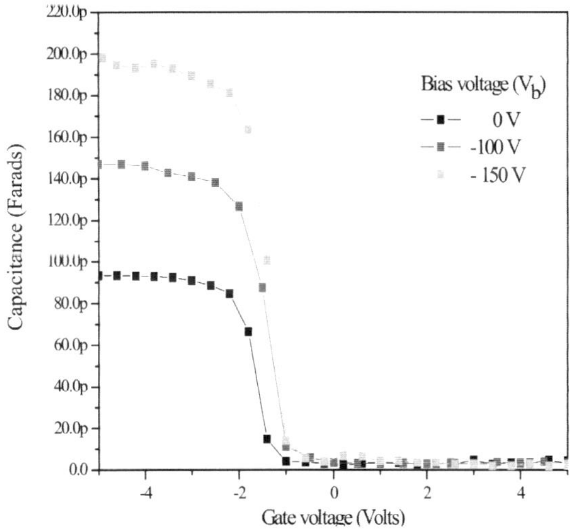

Figure 1. Raman spectra of TiO_2 thin films on Si. **Figure 2.** C-V curves of Al/TiO_2/Si capacitors.

From fig. 2 it can be seen that the capacitance in the accumulation region was found to be increased with increase of substrate bias voltage from 0 to -150 V. The possible explanation for the enhancement in the accumulation capacitance of high substrate bias voltage is the reduction in the TiO_2 thickness from 65 to 54 nm; it could be attributed to the antisputtering phenomenon at higher substrate bias voltage by the collision of highly accelerated positive Ti ions [8]. Besides, large depletion region of a C-V curve leads to large interface defects. It is clear from fig. 2 that the depletion layer width was decreased noticeably with substrate bias voltage resulting improvement in the interface quality at TiO_2/Si stack [9]. Eventually, The dielectric constant calculated from the measured C-V characteristics of the un biased films was 9.8 and it increased to 19.6 for the films deposited at -150 V due to decrease in structural defects can enhance the dielectric constant.

Atomic force microscope (AFM) images of TiO_2 films deposited at different substrate bias voltages are shown in fig. 3 (a)–(c). The morphology of the films significantly varied with substrate bias voltage. As the substrate bias voltage increased from 0 to -150 V, the root mean square roughness was increased from 1.538 to 2.739 nm. The reason may be due to high kinetic energy of Ti^+ ions accelerated at high negative substrate bias voltages.

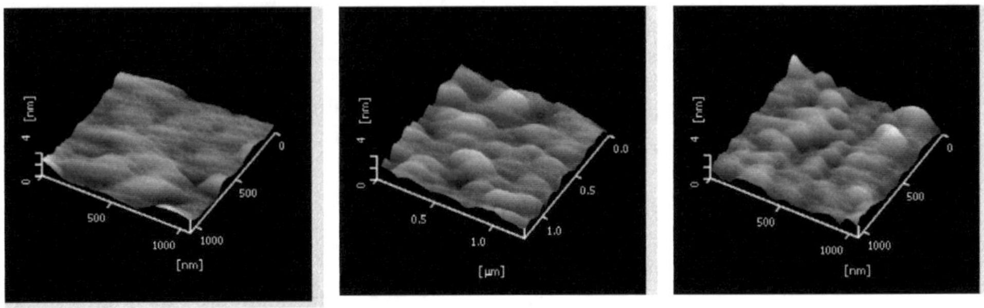

(a) (b) (c)

Figure 3. Atomic force images of TiO_2 films deposited on Si at (a) 0 V, (b) -100 V and (c) -150 V.

Fig. 4 shows the leakage current density versus electric field (J–E) characteristics of the Al/TiO_2/p-Si capacitors. It can be observed that the current density decreases with the substrate bias voltage. The leakage current density measured for un-biased films at gate bias voltage of 2 V was 1.41x 10^{-6} A/cm^2 and it decreased to 1.57x10^{-7} A/cm^2 for films deposited at -150 V. The general conduction mechanisms invoked to explain the current transport in TiO_2 films were Schottky Emission (SE), Poole-Frenkel (PF) emission and Fowler- Nordheim (FN) tunneling. The SE process occurred between a metal and insulating film. The SE conduction was given by the following equation,

$$J_{SE} = A^*T^2 \exp[- q(\Phi_B - \sqrt{qE/4\pi\varepsilon_0 kr}) / kT] \quad (1)$$

Where $A^* = (4\pi qm^*k^2/ h^3)$, the Richardson constant. J_{SE} the current density governed by Schottky emission, m^* the effective mass of electron. Φ_B, k_r, k, and h the Schottky barrier height, dynamic dielectric constant, Boltzmann's constant and Planck's constant respectively.

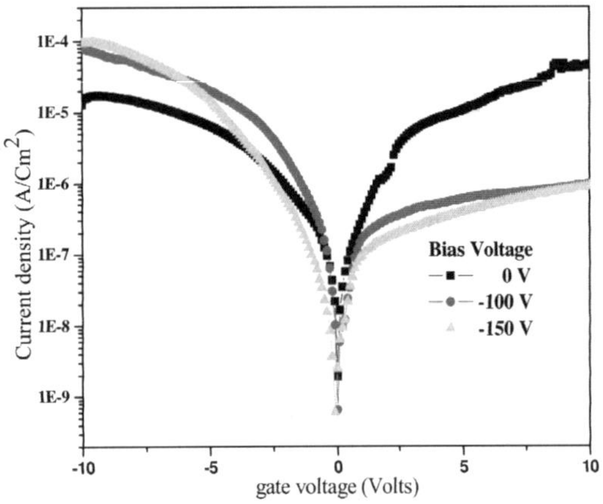

Figure 4. J-E curves of Al/TiO_2/Si capacitors.

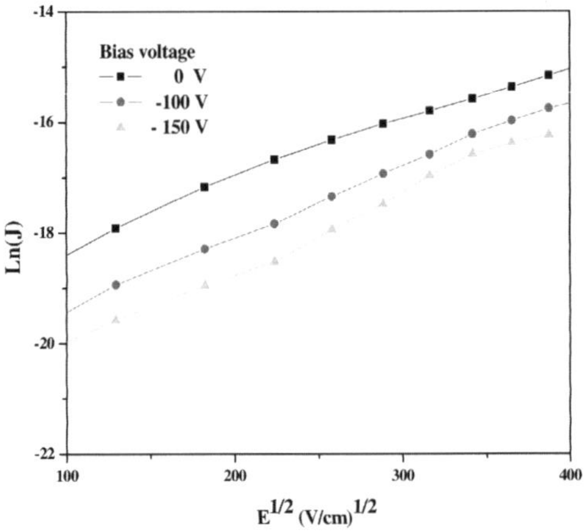

Figure 5. Plots of ln(J) Vs $E^{1/2}$ (SE-Mechanism).

From fig. 5, the ln (J) showed a linear dependence on $E^{1/2}$ for TiO_2 films deposited at different substrate bias voltages from 0 to -150 V consistent with the Schottky conduction mechanism at lower electric fields, where the charge conduction is dominated by the interface between the TiO_2 and Si substrate. Generally, at higher electric fields, two conduction mechanisms were expected: one is Poole–Frenkel and the second is Fowler-Nordheim. The PF emission is due to the emission of trapped electrons into the conduction band of insulator and the FN tunnelling due to the narrowed oxide energy barrier width is also considered as a possible current transport mechanism [10]. Both mechanisms were examined for all the capacitors at higher electric fields. PF conduction mechanism was observed for un-biased films i.e., the plot of ln (J/E) versus $E^{1/2}$ shows linearity as shown in fig. 6. However the films deposited at -100 and -150 V showed Fowler-Nordhiem tunneling conduction mechanism as shown in fig. 7. The reason may be that the vacancies are filled at higher substrate bias voltages and reduce the trapped electron emission [11].

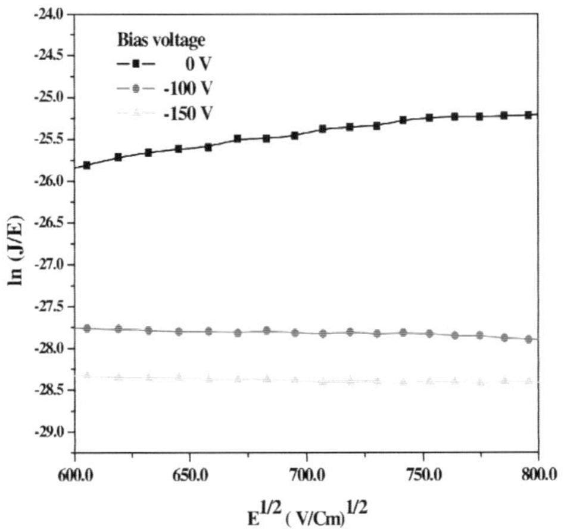

Figure 6. Plots of ln(J/E) Vs $E^{1/2}$ (PF-Mechanism).

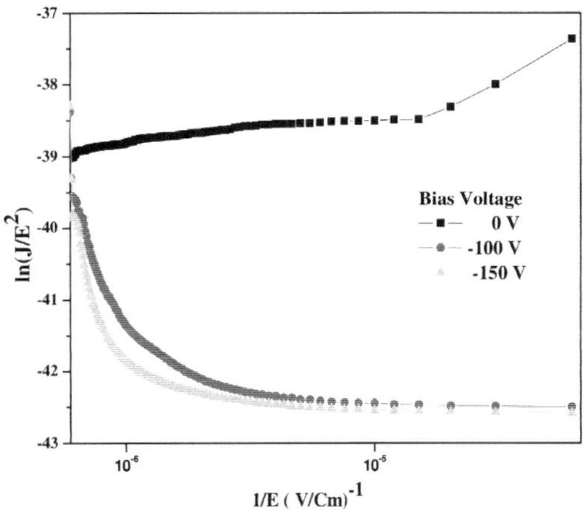

Figure 7. Plots of ln (J/E^2) Vs 1/E (FN-Mechanism).

4. Conclusion

In conclusion, the influence of the substrate bias voltage on structural, dielectric and electrical properties of TiO_2 films was systematically studied. From Raman analysis it was found that the crystallanity was improved with the substrate bias voltage. The dielectric constant was increased with increase of the substrate bias voltage whereas the leakage current was decreased. The Schottky emission conduction mechanism was observed for all the films at low electric fields and at high electric fields the unbiased films exhibit Poole-Frenkel conduction mechanism and the films deposited at -100 V and -150 V shows Fowler-Nordheim conduction mechanism.

Acknowledgement

This work was carried out with the financial support of UGC, New Delhi, India through a sanctioned Major Research Project: F. No. 34-36/2008 (SR) dated: 30-12- 2008. One of the authors P. Kondaiah, thankful to UGC for providing financial support through sanctioned UGC-RFSMS.

References

[1] Thompson S, Packan P and Bohr M 1998 *Intel Technol.J.* **Q3**
[2] Wang M T, Wang T H and Lee J Y 2005 *Microelectronics Rel.* **45** 969
[3] Jagadeesh Chandra S V, Kim J S, Moon K W and Choi C J 2011 *Microelectron. Eng.* doi:10.1016/j.mee.2011.03.156
[4] Kim H S, Campbell S A, Gilmer D C, Kaushik V, Conner J, Prabhu L and Anderson A 1999 *J. Appl. Phys.* **85** 3278
[5] Maiti C K, Samanta S K, Dalapati G K, Nandi S K, Chatterjee S 2004 *Microelectron. Eng.* **72** 253
[6] Sekiya T, Ohta S, Kamei S, Mesot J, Sato T, Takahashi T, Norman M, Randeria M, Kadowaki K, Hinks D 2001 *J. Phys. Chem. Solids* **62** 35
[7] Karunagaran B, Kim K, Mangalaraj D, Yi J and Velumani S 2005 *SolarEenergy Materials & Solar Cells* **88** 199
[8] Huang A P, Xu S L, Zhu M K, Wang B, Yan H and Liu T 2003 *Appl. Phys. Lett.* **83** 3278
[9] Cho B O, Wang J, Sha L and Chang J P 2002 *Appl. Phys. Lett.* **80** 1052
[10] Sze S M, 2007 *Physics of Semiconductor Devices, 3rd Edn.*, Wiley, New York
[11] Chakraborty S, Bera M K, Bhattacharya S and Maiti C K 2005 *Microelectron. Eng.* **81** 188

Institute of Physics
Dirac House, Temple Back
Bristol BS1 6BE UK

CURRAN ASSOCIATES INC
proceedings
.com